Molecular Biology Intelligence Unit

Regulation of the RAS Signaling Network

Hiroshi Maruta
Antony W. Burgess
Ludwig Institute for Cancer Research
Royal Melbourne Hospital
Victoria, Australia

CHAPMAN & HALL
ITP An International Thomson Publishing Company
New York • Albany • Bonn • Boston • Cincinnati • Detroit • London • Madrid • Melbourne •
Mexico City • Pacific Grove • Paris • San Francisco • Singapore • Tokyo • Toronto • Washington

R.G. Landes Company
Austin

Molecular Biology Intelligence Unit
REGULATION OF THE RAS SIGNALING NETWORK

R.G. LANDES COMPANY
Austin, Texas, U.S.A.

Printed in the U.S.A.

Please address all inquiries to the Publishers:
R.G. Landes Company, 909 Pine Street, Georgetown, Texas, U.S.A. 78626
Phone: 512/ 863 7762; FAX: 512/ 863 0081

North American distributor:
Chapman & Hall, 115 Fifth Avenue, New York, New York, U.S.A. 10003

CHAPMAN & HALL

U.S. and Canada ISBN: 0-412-11861-0

Library of Congress Cataloging-in-Publication Data

Regulation of the RAS signaling network / [edited by] Hiroshi Maruta, A.W. Burgess
p. cm. — (Molecular biology intelligence unit)
Includes bibliographical references and index.
ISBN 0-57059-404-X (RGL: alk. paper); ISBN 0-412-11861-0
1. Ras oncogenes. 2. Cellular signal transduction I. Maruta, Hiroshi. II. Burgess, Antony, 1946- . III. Series.
RC268.44.R37R44 1996
616.99'4071—dc20

96-35747
CIP

Publisher's Note

R.G. Landes Company publishes six book series: *Medical Intelligence Unit, Molecular Biology Intelligence Unit, Neuroscience Intelligence Unit, Tissue Engineering Intelligence Unit, Biotechnology Intelligence Unit* and *Environmental Intelligence Unit.* The authors of our books are acknowledged leaders in their fields and the topics are unique. Almost without exception, no other similar books exist on these topics.

Our goal is to publish books in important and rapidly changing areas of bioscience and environment for sophisticated researchers and clinicians. To achieve this goal, we have accelerated our publishing program to conform to the fast pace in which information grows in bioscience. Most of our books are published within 90 to 120 days of receipt of the manuscript. We would like to thank our readers for their continuing interest and welcome any comments or suggestions they may have for future books.

Shyamali Ghosh
Publications Director
R.G. Landes Company

In Memory of Irving Sigal

Irving Sigal (1953-1988)

Much of the study of the molecular biology of RAS—in particular, the structure-function work—was initiated by the late Irving Sigal (1953-1988), when he joined Ed Scolnick's group at Merck, Sharp & Dohme Research Laboratories in 1983. Irving was a chemist by training. Using site-directed mutagenesis to create a series of RAS mutants, he and his colleagues identified the unique domain of RAS which is required for its interaction with downstream molecules (for further details, see chapter 5). This section of RAS is now called the "effector" or "switch I" domain. The "effector" domain mutants of RAS, many of which are no longer oncogenic, have served as useful tools for distinguishing between the mode of interaction between RAS and many RAS-binding proteins. Irving was also actively involved in both the purification and cloning of the first RAS effector protein, called GAP1, a RAS GAP of 120 kDa (see chapter 5). If Irving were still alive, it is most likely that he would have joined us in editing this first book on RAS.

Unfortunately, soon after the cloning of GAP-1 was completed PanAm flight 103, carrying 270 people including Irving, was downed over Lockerbie in England by an act of terrorism on December 21, 1988. There were, sadly, no survivors of this tragic air crash. It was a great loss not only for his own family, in particular his wife Cathy Sigal, but also for us and the whole RAS research community. Thus, we would like to dedicate our first RAS book to the late Irving Sigal. Irving was a great seeder of the RAS field, and, as will be seen in this book, the world is now harvesting fruit from the rich plantation that he left us. We hope that this book will also stimulate others to creative efforts in the field of RAS research, bringing new fruits to both this and coming generations...

CONTENTS

EDITORS

Hiroshi Maruta
Ludwig Institute for Cancer Research
Royal Melbourne Hospital
Victoria, Australia
Chapter 5

Antony W. Burgess
Ludwig Institute for Cancer Research
Royal Melbourne Hospital
Melbourne, Australia
Chapter 3

CONTRIBUTORS

David D.L. Bowtell
Peter MacCallum Cancer Institute
Melbourne, Australia
Chapter 1

Min Han
Department of Molecular, Cellular and Developmental Biology
University of Colorado
Boulder, Colorado, U.S.A.
Chapter 2

Toshihisa Ohtsuka
Department of Molecular Biology and Biochemistry
Osaka University Medical School
Suita, Japan
Chapter 6

Isabel Sattler
Department of Microbiology and Molecular Genetics
University of California
Los Angeles, California, U.S.A.
Chapter 4

Kazuya Shimizu
Department of Molecular Biology and Biochemistry
Osaka University Medical School
Suita, Japan
Chapter 6

Meera Sundaram
Department of Molecular, Cellular and Developmental Biology
University of Colorado
Boulder, Colorado, U.S.A.
Chapter 2

Yoshimi Takai
Department of Molecular Biology and Biochemistry
Osaka University Medical School
Suita, Japan
Chapter 6

Fuyuhiko Tamanoi
Department of Microbiology and Molecular Genetics
University of California
Los Angeles, California, U.S.A.
Chapter 4

PREFACE

Cancer research has revolutionized our understanding of mammalian cell biology, and in many respects the discoveries associated with the RAS protein have been pivotal to our spectacular progress. A little over thirty years ago, Jennifer Harvey discovered that a new sarcoma virus had been produced by passaging the murine Moloney Leukemia virus in rats (Harvey, JJ, Nature 204:1104, 1964). The new virus (Ha-MSV) rapidly induced sarcomas near the site of injection of the virus. Eventually, Edward Scolnick and his team at the National Cancer Institute in Bethesda determined that Ha-MSV contained a large insert of rat genetic material between the Moloney helper virus sequences (Scolnick EM and Parks WP, J Virol 13:1211, 1974). Clearly, the rat sequences were critical for the acute-transforming properties of the new virus (Shih TY, Weeks MO, Young HA and Scolnick EM, J Virol 31:546, 1979). Soon after the RAS studies were published, Michael Bishop, Harold Varmus and Dominique Stehelin determined that another viral oncogene, *Src*, had also been captured from its host cells (Stehelin E et al, Nature 260:170, 1976).

Although the link between human cancers and the oncogenic animal viruses had been difficult to detect, Robert Weinberg and his colleagues believed that the genetic information associated with human cancers would also be capable of transforming normal cells. They adapted a DNA transfection technique to detect the cancerous gene(s) associated with human tumors by focus formation in a mouse fibroblast cell line (Shih C, Shilo BZ, Goldfarb MP, Dannenberg A and Weinberg RA. Proc Natl Acad Sci 76:5714, 1979). Almost simultaneously three independent groups determined that the human oncogene identified by Weinberg's team corresponded to the rat gene (RAS) responsible for the transforming properties of the Harvey and Kirsten sarcoma viruses (Shih et al, Nature 290:261, 1981; Krontiris TG and Cooper GM Proc Natl Acad Sci 78:1181C, 1981; Perucho M et al Cell 27:467, 1981). Clearly, some of the mutations responsible for human cancer operated by mechanisms similar to the oncogenes carried by cancer-causing viruses of animals.

As well as identifying a myriad of possible mutations which can contribute to the cancerous state, work over the last fifteen years has also confirmed the commonality of many oncogenic events (e.g., p53 and RAS). Despite the enormous diversity of this research, the centrality of the RAS oncogene and the c-RAS protein in cellular proliferative and differentiative processes has remained.

The initial observations indicated that the oncogenic RAS and normal human RAS proteins differed by a single amino acid. Scolnick's group had observed that the viral RAS (v-RAS) protein was phosphorylated and was capable of binding GTP (Shih TY et al, J. Virol 42:253, 1982). Although normal forms of human RAS bind GTP with equal affinity, normal RAS hydrolyzes the GTP to GDP at a much higher rate than the oncogenic RAS (McGrath et al, Nature 310:644, 1984). More importantly, the intrinsic GTPase activity of only normal RAS, and not oncogenic RAS, is highly stimulated by GTPase activating proteins (GAPs) (Trahey M and McCormick F, Science 238:542, 1987). By analogy with other G-protein signaling systems, it was suggested that only the GTP-bound form of RAS acted to signal mitogenic, differentiative or motility processes. It is now clear that many cellular regulatory systems act by controlling the concentration of membrane-associated GTP-RAS.

The importance of RAS continues to grow as the biochemistry of more and more cytokine/growth factor/transforming systems are unraveled. We now have exquisite details of the three-dimensional structure of RAS (Franken SM et al, Biochemistry 32:8411, 1993; Milburn MV et al, Science 297:939, 1990), and there is even a high resolution crystal structure of the RAS/Raf interface (Nassar N et al, Nature 375:1554, 1995). It is clear that activated RAS controls several critical processes, but whilst the roles of these processes in normal cell division or the maintenance of the oncogenic state remain to be determined, oncogenic RAS is clearly an important target for the development of more specific and potent cancer therapies (James GL et al, Science 260:1937, 1993;), (Sepp-Lorenzino L et al, Cancer Res 55:5302, 1995).

Tony Burgess
Hiroshi Maruta
May, 1996

CHAPTER 1

GENETICS OF RAS SIGNALING IN *DROSOPHILA*

David D.L. Bowtell

I. INTRODUCTION

As in other multicellular organisms, RAS proteins play a key role in the signaling pathways that regulate cell growth and differentiation in *Drosophila*. Although these proteins were first identified in *Drosophila* using cross-species hybridization, it has been the use of classical genetics which has overwhelmingly driven investigations of RAS function in *Drosophila*. This focus has been successful in placing RAS in complex signaling pathways for several reasons: genetic screens generally make no assumptions about the biochemical activity of the components of the signaling pathway, thereby facilitating the isolation of proteins with novel activities; a *functional* requirement for the protein in the pathway is established at the outset, which is often lacking when proteins are isolated by other means; and tests of cell autonomy and epistasis frequently position the requirement for the protein in the pathway, even before its biochemical activity is known.

Although RAS proteins are involved in many facets of *Drosophila* development, this review will focus mainly on their role in the signaling pathways downstream of three tyrosine kinase receptors; sevenless, the *Drosophila* EGF receptor homologue (D-EGFR) and torso. Sevenless has a very restricted role in development, being specifically required for differentiation of the R7 photoreceptor cell in the developing eye. Torso is required in several sites but its role in the development of the terminal structures of the embryo is best characterized. The D-EGFR is required in many tissues, including the development of the embryonic ventro-lateral midline cells, the adult cuticle, wings and photoreceptor cells.

Regulation of the RAS Signaling Network, edited by Hiroshi Maruta and Antony W. Burgess.

To assist the reader, background information about the development of these tissues and some of the genetic analyses used in these systems is provided at the outset. The review is then organized "top-down," beginning with the receptors and their ligands. Included in this section is a discussion of a number of genes which appear to modulate the activity of the D-EGFR. Section IV deals with membrane proximal signaling proteins, including RAS itself, and proteins which either directly modulate RAS activity or in other ways provide a link between RAS and receptors. Section V describes a serine threonine kinase cascade downstream of RAS which is largely common to all three receptors. The final section describes how this kinase cascade leads to a direct biochemical link to a number of nuclear proteins and discusses other nuclear proteins essential for RAS signaling but for which a biochemical function is not yet defined.

II. BACKGROUND TO THE EXPERIMENTAL SYSTEMS

The Developing Eye

A brief outline of the development of the *Drosophila* eye is necessary for a discussion of the sevenless and D-EGFR signaling pathways. A more detailed account can be found in several excellent reviews of *Drosophila* eye development.[1-4]

The adult *Drosophila* eye consists of a compound of approximately 800 hexagonal individual units or ommatidia (Fig. 1.1A). Each ommatidium is in essence an individual eye and consists of eight photoreceptor cells, an overlying lens secreted by four cone cells, bristle cells and a sheath of pigment cells which optically insulates the ommatidium. The photoreceptors are of three types (R1-R6, R7 and R8) based on their morphology, spectral analysis of opsins and a large amount of molecular and genetic data. They have specialized membranes, rhabdomeres, which concentrate the photosensitive opsins.

The eye and antenna develop from a flat disk of epithelium, the eye-antennal imaginal disk (Fig. 1.1B). Imaginal disks are comprised of clusters of cells that are sequestered during embryogenesis and will eventually form most of the adult epidermal structures, including legs, antenna and eyes. Differentiation and pattern formation in the developing eye begins in the third (last) larval instar and continues during pupation. It commences at the posterior margin of the eye disk (right, panel B) and spreads in a wave to the anterior margin of the disk over a 48 hour period. The boundary between unpatterned (anterior) and differentiating (posterior) epithelium is marked by a depression, the morphogenetic furrow. 'Rosettes' of cells form posterior to the furrow and these will comprise the future ommatidia (Fig. 1.1C). Cells within these rosettes differentiate in well characterized sequence beginning with the R8 cell, followed pair wise by R2 and R5, R3 and R4, R1 and R6

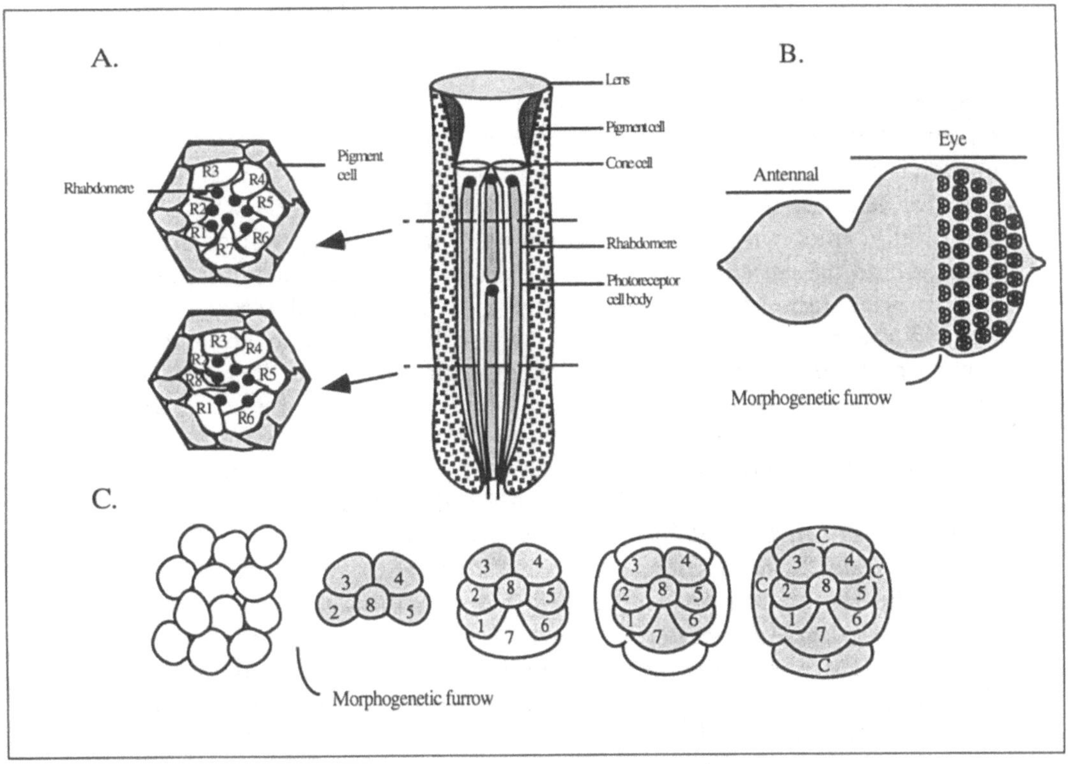

Fig. 1.1. (A) The adult eye consists of a compound array of hexagonal units or ommatidia. It develops during the third larval instar and pupation from the eye antennal imaginal disk (B) in a well defined sequence (C, see text).

and finally by R7 (Fig. 1.1C). Two cells of unknown developmental fate called the mystery cells (not shown) display some characteristics of early photoreceptor differentiation, but then fail to develop into photoreceptors. These cells, which appear to represent a prephotoreceptor, frequently differentiate into photoreceptors in certain mutants and provide additional information about mutant phenotypes. Differentiation of the cone and pigment cells follows that of the photoreceptors. Importantly, differentiation of the photoreceptors occurs in response to local environmental cues elaborated by adjacent differentiating cells and is strictly lineage-independent. A number of mutations have been identified in which the developmental sequence is interrupted at specific points, suggesting that as cells differentiate they instruct later appearing cells of their developmental fate. Hence, failure early in the differentiation sequence usually leads to subsequent errors in cell fate choice. Conversely there are mutations in genes such as *sevenless*, which is required at the end of the developmental sequence, where photoreceptor development is normal until the addition of the R7 photoreceptor cell.

Development of the Embryonic Terminal Structures

The *Drosophila* embryo develops rapidly from a fertilized egg to a first stage (instar) larva over a period of approximately 24 hours. Cytoplasmic transfer experiments and, more recently, saturation genetic screens define four major programs of embryonic axis and pattern formation—anterior, posterior, dorso-ventral and terminal. The development of terminal structures has been reviewed in detail elsewhere.[5,6]

Briefly, the terminal system represents a range of specialized structures including cuticular structures at the anterior (acron) and posterior (telson) ends of the embryo (Fig. 1.2). Mutational screens have defined the terminal class of genes, whose loss results in absence of some or all of these structures. Recognition that these loci are the major controllers of these embryonic structures provided the starting point for a molecular investigation of this developmental system. Mutational analysis defined seven genes which must be expressed maternally for the development of the anterior and posterior terminal structures; *fs(1)Nasrat, fs(1)polehole, torsolike, torso, trunk, l(1)polehole* (D-Raf) and *corkscrew*. Loss of these genes does not result in cell death, but rather the respecification of the terminal structures to a subterminal and central developmental fate. Only *torso, torsolike* and *trunk* give a strict terminal phenotype, as the remainder are also involved in other developmental processes in the embryo. Importantly, of the seven genes only *torsolike* is required in the follicle cells, indicating that torsolike provides an inductive signal for development of the embryo adjacent to these cells.

Genetic Screens

The sevenless, torso and D-EGFR receptors were initially identified in genetic screens for mutations affecting photoreceptor differentiation or embryonic development. Subsequent saturating screens for recessive mutations in these developing tissues identified relatively few of the other components in the pathways downstream of these receptors. Mutations in components which operate late in development may remain hidden from this analysis, as recessive mutations in common signaling components can result in embryonic lethality before the development of tissues such as the eye has commenced. Several strategies have been adopted to overcome this problem.

Enhancer traps

One strategy to discover genes involved in the development of a particular tissue is enhancer trapping. The basis of this technique is to use a transposable element which carries a reporter gene (e.g., lacZ) that is sensitive to activation by regulatory elements in the vicinity of the integration of the element in the genome. The most commonly used transposable element for this purpose is the P-element, which can be conditionally mobilized to create thousands of transgenic flies

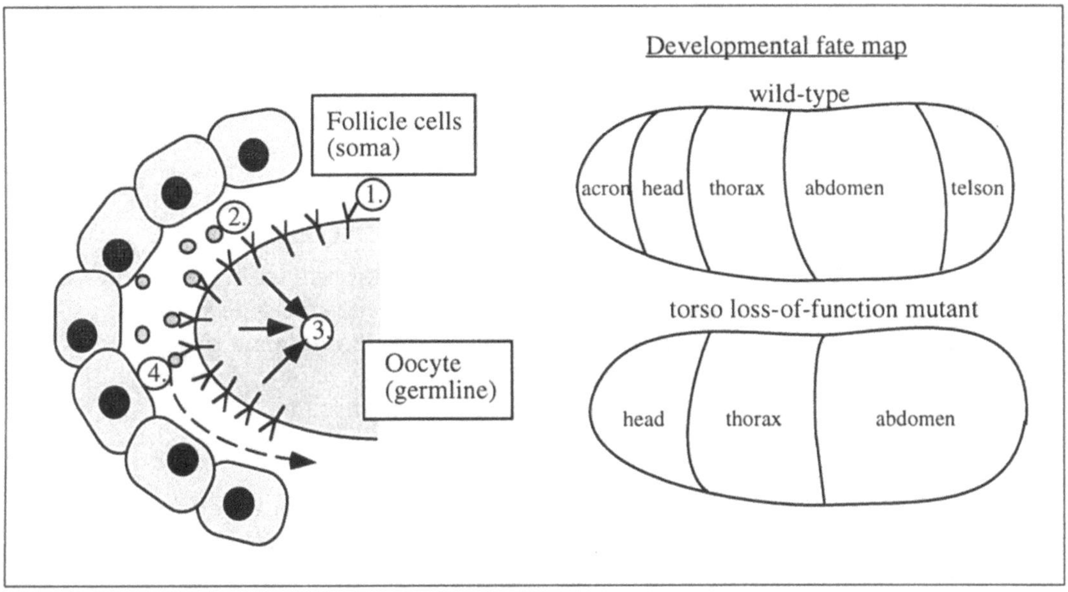

Fig. 1.2. A large amount of genetic and biochemical data indicate that the torso protein (1) receives a localized inductive signal (2) from terminally located follicle cells to trigger a signaling pathway (3) which directs expression of genes required for embryonic terminal differentiation. Other genes, including D-sor *(D-MEK),* tailless*, and* huckebein*, have been identified more recently which are required downstream of torso and add to the initial complement of seven terminal class genes in this signaling pathway. (4) Expression of the receptor appears to also be important in limiting the range of action of the ligand.*

in which the element is inserted at discrete sites within the genome. The presence of the element adjacent to a gene of interest can be identified by staining the progeny of these animals for β-galactosidase activity. This approach works particularly well in tissues where it is possible to easily gain precise information about single cell patterns of expression, such as in the embryo or in the developing eye. Once a gene of interest is identified the element can be conditionally mobilized once more, frequently resulting in a local reinsertion or an imprecise excision that creates mutants in the gene of interest for further study.

Mosaic analysis

The creation of somatic mosaics is a particularly powerful approach for studying genes whose mutation results in embryonic lethality, but which are also required at later stages. This technique is widely used to investigate the differentiation of cells in the eye, or in embryos where the maternal gene contributes to early embryonic development. Traditionally, mosaic analysis (also known as clonal analysis) has relied on creating a cross in which flies carry a heterozygous null mutation in the gene of interest on one chromosome. An innocuous phenotypic marker reports on the genotype of the cell. Mitotic recombination is induced by X-irradiation during development to create two daughter cells, one homozygous mutant lacking the marker and one homozygous wild type carrying the marker. As the mutant cell divides it produces a clone whose phenotype reflects loss of the gene of interest. Importantly, the clones are typically small and do not effect the viability of the animal. The wild type sister cell also produces a clone, called the twin spot, which can be a valuable control. A common marker in the eye is the *white* gene. Homozygous mutant cells lack pigment and are readily recognizable against the pigmented wild type cells. A recent variant of this approach utilizes an inducible recombinase to generate at high frequency[7,7a] clones which are marked by absence of a Myc-epitope tagged protein expressed from the homologous chromosome.

Modifier screens

Of great recent significance has been the use of modifier screens to uncover many signaling components. The basis of these screens is to create circumstances where a given signaling pathway is very sensitive to small increase or partial decrease in any upstream or downstream component of the pathway. The strategy for this is outlined schematically in Figure 1.3.

Variations on the approach shown in Figure 1.3 include the isolation of mutations which can compensate for the phenotypic effect of a receptor (or any other pathway component) whose activity is below a critical threshold for cell differentiation. Such mutations are referred to as 'suppressors'. Gain-of-function mutations in factors that are positively required for signaling, or loss-of-function mutations in negative regulators, are uncovered in these screens. Another approach is to suppress the effects of a weakly dominant form of a receptor, by a heterozygous loss-of-function mutation in a component required downstream for signaling.

In each of these examples, it is only the cell bearing the sensitized pathway that responds to the effects of these heterozygous mutations, even though the component may operate in many cell types. This serves to simultaneously limit the effects of the mutation (e.g., lethality) and to associate the mutation with the pathway in question. Experiments utilizing these modifier screens appear throughout this review.

III. PROTEIN TYROSINE KINASE RECEPTORS

Sevenless is Specifically Required for R7 Photoreceptor Cell Formation

The R7 photoreceptor makes a UV sensitive opsin that allows flies to see light in this range of the spectrum. This provides a convenient behavioral test for absence of the R7 photoreceptor and allowed the isolation of the *sevenless* mutation.[9] The only apparent defect in *sevenless*

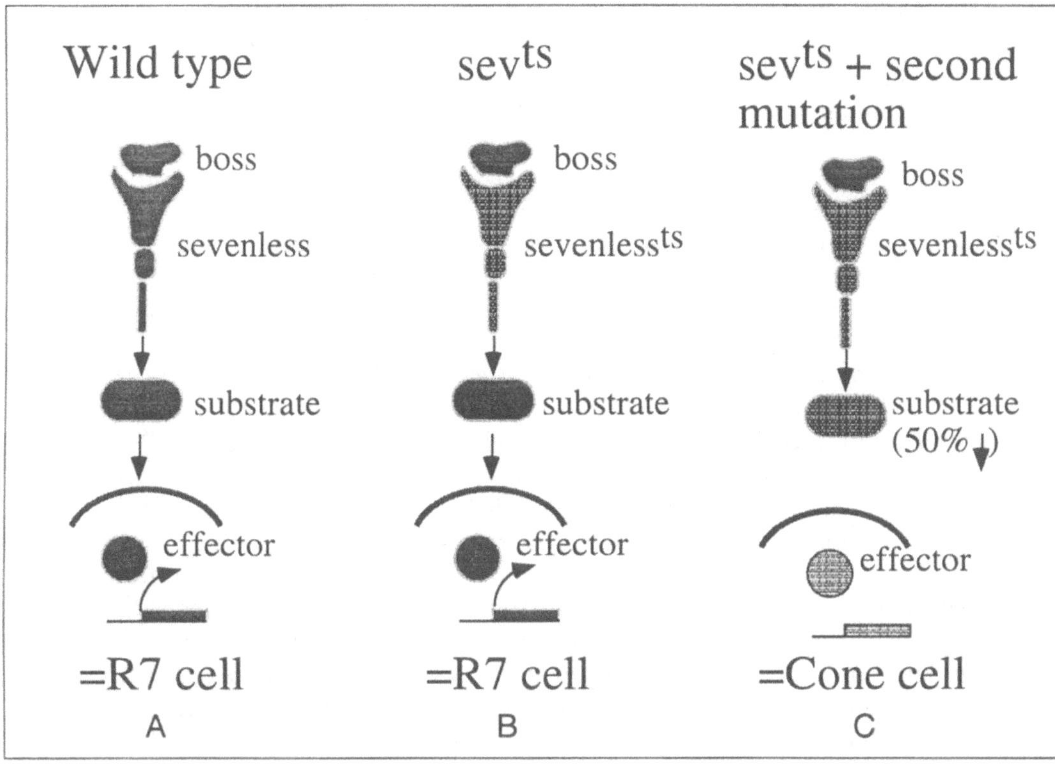

Fig. 1.3. A modifier screen for proteins downstream of the sevenless receptor.[8] In this experiment heterozygous mutations in components positively required for signaling in the pathway are detected in an F1 screen. Mutations isolated from this screen are referred to as Enhancers of sevenless E(sev). (A) In this example the R7 cell requires a highly specific receptor (sevenless) for its differentiation but the receptor utilizes, in part, rather general downstream signaling components. (B) Circumstances are created where the precursors of the R7 cell carry a barely functional receptor and will differentiate normally when all the other signaling components are fully functional. (C) However, when there is a 50% decrease in a upstream or downstream component the balance tips towards a failure of cell differentiation and this phenotype can be scored (formation of a cone cell) and the mutation recovered.

flies is the absence of the R7 photoreceptor, which adopts a nonneuronal cone cell fate.[10] Mosaic analysis demonstrated that sevenless is required only in the precursor of the R7 photoreceptor.[11-13] Subsequent cloning and expression analysis of *sevenless* showed it to code for a large tyrosine kinase receptor which is expressed on a subset of developing photoreceptor cells, including the R7 photoreceptor, and cone cells.[14-19] The cell specific pattern of sevenless expression changes rapidly during eye development. However, by using a temperature sensitive allele of *sevenless* to define the period when sevenless is required[20] and by replacing the intricate pattern of sevenless expression with a general one[21,22] it became clear that sevenless is only required for a brief period in the R7 precursor and that its expression elsewhere is immaterial. Expression of a constitutively active allele of *sevenless* leads to transdetermination of cone cells into R7 cells[23] but does not affect the development of the other photoreceptor cells. This finding, together with the fact that the R7 precursor becomes a cone cell in the absence of sevenless, indicates that the cone cells and R7 precursor form an equivalence group in terms of their differentiation potential.[24] If so, then why don't the sevenless expressing cone cells become R7 cells? At least part of the answer is that sevenless activity is restricted by the limited distribution of a cell surface associated ligand called boss.

Boss

Bride of sevenless (boss) is an unusual ligand for a receptor, as it is an 896 amino acid seven-transmembrane protein with a large (498 amino acid) extracellular domain, which interacts with sevenless.[25,26] Boss is expressed specifically in the developing R8 cell, which is immediately adjacent to the developing R7 cell[26,27] (see also Fig. 1.1C above). As anticipated from its primary sequence, boss is not freely diffusible. An interesting consequence of this is that boss and sevenless accumulate where the membranes of the R7 and R8 cells contact each other.[15,28] Sevenless is required for the apical localization of boss in the developing eye disk.[28] Both phenomena probably reflect a partially immobilized receptor-ligand interaction. This cell-cell interaction can be reproduced in vitro by mixing cells which express either sevenless or boss.[27] Under these circumstances cell aggregates are formed and phosphorylation of sevenless is stimulated.[29] Surprisingly, boss is translocated from the cell surface of boss-expressing cells to the cytoplasm of cells expressing sevenless in this assay.[27] This also occurs in vivo, where the entire boss protein is transferred from the R8 cell to large multivesicular bodies within the R7 cell, in a sevenless-dependent manner.[27,28] Although receptor mediated internalization of ligand has precedents in other systems, this example is unusual in that the ligand is a membrane-associated protein. Given the structure of boss there is the opportunity for signals to be sent in both directions, that is for it to be both a ligand *and* a receptor. Although at present there is no direct evidence of this, it is noteworthy that both the extracellular domain and the seven transmembrane domain are necessary for boss-dependent phosphorylation of sevenless, in the above in vitro cell system.[29] An extracellular domain fragment acts as an antagonist in this system, perhaps because activation of sevenless may require a high affinity interaction or needs to be oligomerized and this can only be provided by the membrane bound protein. The tight restriction on the cells in which boss is expressed (R8 cell only) and the fact that it remains membrane-associated must serve, in part, to limit the cells in which sevenless is activated, as the cone cells (which normally express sevenless) do not contact the R8 cell. That ectopic expression of boss results in cone cells adopting an R7 cell fate is entirely consistent with this view.[30]

However, in wild type flies the developing R3 and R4 cells, and to lesser extent the R1 and R6 cells, express the sevenless protein and contact the R8 boss-expressing cell but do not respond to sevenless.[15] This anomaly may be explained by their earlier specification in the developmental sequence, so that they are refractory to the sevenless signal. Support for this idea comes from flies mutant for the genes *rough* and *seven-up*. In *rough* flies the developmental sequence for the R1-R6 cells is interrupted at the point when R3 and R4 differentiate[31] and more cells appear to adopt an R7 fate than in wild type animals.[30] This is consistent with the notion that the prior specification of the R1-R6 cells in wild type flies limits their ability to respond to the late appearing signal from sevenless. Similarly, in the absence of seven-up the R1, R3, R4 and R6 cells form supernumary R7 cells (see section VI). The way in which the R1-R6 cells process sevenless and boss also differs from the R7 precursor. Whereas the R7 cell internalizes the boss ligand in association with the sevenless receptor, the R1, R3, R4 and R6 cells fail to do so.[32] Although this is suggestive of a link between the ability of cells to internalize boss and their ability to form R7 cells, internalization cannot be essential for signaling by sevenless, as a constitutively active sevenless protein can direct the formation of R7 cells in a boss-independent manner.[23]

Torso, a Receptor Tyrosine Kinase Required for Development of the Terminal Embryo Structures

As described above, the terminal group represents seven maternally required genes necessary for proper development of the terminal structures of the embryo (reviewed in ref. 5). Of these *torso* is unique in having both loss-of-function[33] and gain-of-function alleles[34,35] whose phenotypes are opposite. The availability of both types of mutations has been important in defining the role of torso. In embryos which lack maternally derived torso the terminal structures are absent and they are respecified to a subterminal and central embryonic fate, resulting in expanded head and abdominal regions. Strong gain-of-function alleles result in the subterminal and central regions of the embryo being respecified to a terminal fate. This results in a "fused embryo" in which the terminal regions are enlarged and segmentation of the thorax and abdomen is suppressed (see Fig. 1.2). Dominant *torso* alleles have been used to demonstrate that of the terminal class of genes, only *corkscrew* and D-Raf (*l(1)polehole*) are required downstream of *torso*.[36-39] (Other genes which are also downstream of *torso*, but were not identified in this initial analysis, are described in sections IV-VI).

Torso is a 923 amino acid protein with all the features of a tyrosine kinase receptor.[40] The kinase domain is of the "kinase insert" type seen in the mammalian PDGF, CSF-1 and C-kit receptors, in which the kinase domain is interrupted by a short divergent peptide sequence. The protein is localized to the surface of the developing embryo and, importantly, is not restricted to the terminal regions of the embryo.[36] Biochemical analyses of *torso* mutant embryos demonstrate that the wild type protein has intrinsic autophosphorylation activity. This is enhanced in strong gain-of-function alleles and absent in several loss-of-function alleles which have point mutations in conserved residues within the kinase domain.[41] The behavior of the gain-of-function alleles and the generalized expression of the wild type protein indicate that torso activity must be normally restricted to the terminal regions of the embryo. This is achieved by limited access to the torso ligand.

Torso-Like and Trunk

The product of the *torso-like* gene is the most likely ligand for torso. *Torso-like* mutant embryos have essentially the same phenotype as *torso* loss-of-function mutants and genetically *torso-like* is upstream of *torso*.[33] Mosaic analysis has demonstrated that, in contrast to other terminal class genes, *torso-like* is required only in the follicle cells at the termini of the embryo.[38] The gene has been cloned and is predicted to produce a secreted protein of 356 amino acids,[42,43] which is expressed in a gradient at the ends of the embryo.[42] Overexpression of torso-like produces a phenotype similar to that seen in embryos expressing dominant alleles of *torso*.[42,43] The level of tyrosine phosphorylation of torso, an indicator of its activation, is reduced in embryos mutant for torso-like.[41]

Although torso-like is favored as the ligand for torso, another possible candidate is trunk. The *trunk* gene encodes a predicted secreted protein of approximately 230 amino acids which has some characteristics similar to spazle, the ligand for the toll receptor.[44] As for torso-like, embryos mutant for trunk show reduced levels of torso phosphorylation.[41] Trunk is also required upstream of torso. However, in contrast to torso-like, trunk is required in the developing embryo rather than the adjacent follicle cells.[45] Alternatively trunk, together with fs(1)Nasrat and fs(1) Polehole, may function in regulating the expression or localization of torso-like, as these genes are required upstream from *torso* but do not regulate *torso* expression[38] (see discussion in ref. 6).

There is an interesting reciprocal regulation of the spatial activity of torso and its ligand. That is, the torso ligand restricts the region in which torso is active and the expression of torso itself is important in limiting the range of activity of the ligand. This was shown by comparing the effect of ectopically expressed torso in wild type and torso mutant embryos.[46,47] Whereas ectopic expression of torso in the middle of wild type embryos has no specific effect, in embryos which lack the torso protein there is inappropriate formation of terminal structures in this region. This appears to reflect trapping of the ligand by torso in the region of the termini in wild type embryos, and is reminiscent of the sevenless-dependent localization of boss.[32]

The D-EGFR is Required for the Development of Many Structures in *Drosophila*

The D-EGFR is a receptor tyrosine kinase with a high degree of homology to the mammalian EGF and neu receptors.[48] Mutations have been isolated independently, giving rise to its various names; D-EGFR, Ellipse, torpedo, faint little ball (flb). The latter two correspond to loss-of-function mutations, whereas Ellipse is a weak dominant gain-of-function mutation.[51] Loss of the D-EGFR protein results in a wide range of developmental defects and embryonic lethality, as the protein is required for ventral ectodermal development, malpighian tubule formation, germ band retraction, head development and dorso-ventral patterning of the follicle cells during embryogenesis (see ref. 49 and references therein). A large allelic series of complete and partial loss-of-function mutations has recently been characterized molecularly.[50]

Although null mutations in *D-EGFR* result in embryonic lethality, it has been possible to study its role and its interaction with other genes by creating mutant clones in otherwise normal tissue. In the eye[7,52] and wing,[53] clones of cells homozygous for null alleles of *D-EGFR* are uniformly smaller than control wild type clones, indicating that *D-EGFR* is required for cell proliferation. It is also required for photoreceptor differentiation, as mutant clones in the eye lack photoreceptors, and for normal wing vein formation. The importance of the D-EGFR in the eye and wing has been also highlighted by the use of *Ellipse*.

The weakly dominant allele, *Ellipse*, has been particularly useful in analyzing D-EGFR function in *Drosophila*. Flies which are heterozygous for one wild type copy of *D-EGFR* and one copy of *Ellipse* have a distinctive wing vein phenotype[54,55] and a slight roughening of the exterior of the eye due to disordered photoreceptor development.[51,52] Given that this mosaic analysis has shown that the D-EGFR is required for photoreceptor development, it is surprising that Ellipse flies have *reduced* numbers of photoreceptors. Its role in photoreceptor development is not well understood but it appears that the D-EGFR is initially a repressor of cell determination early in eye development and that later it is required for formation of most, if not all, photoreceptor cells.

Both the wing and eye phenotypes of Ellipse are much more severe in homozygotes. Importantly, the effects of Ellipse can be suppressed when the contribution of the wild type gene is removed. This is done by creating flies which carry *Ellipse* in trans to a deletion of the wild type copy of *D-EGFR*.[51] This shows that the signal from Ellipse is sensitive to the contribution of the wild type D-EGFR protein. Importantly, the activity of Ellipse is also sensitive to the dosage of other components in the pathway. As discussed later, the phenomenon of enhancement and suppression of the Ellipse phenotype has been utilized extensively to identify many of the components in the D-EGFR signaling pathway.

Spitz

The best characterized ligand for the D-EGFR is spitz, which is 230 amino acids in length and has a high degree of homology to mammalian EGF and TGF-α.[56] Interestingly, like other members of the EGF family, spitz contains a signal sequence and single transmembrane domain and has a dibasic sequence which could allow it to be cleaved from a membrane bound form. Recent biochemical analysis of spitz has shown that a genetically engineered secreted form, but not the membrane bound protein, can stimulate in vitro autophosphorylation of the D-EGFR expressed in insect cells.[49] Downstream responses typically seen following EGF stimulation of the mammalian EGF receptor are also elicited; induced Drk binding (the Grb2 homologue) and MAP kinase activation.

Mosaic analysis demonstrates that spitz is absolutely required in the eye for the formation of the R8 cell and has a partial requirement in the developing R2, R5, R3 and R4 photoreceptors.[57,58] These findings are consistent with spitz being a diffusible ligand which acts over a fairly short range within ommatidia and imply that it must be present at the start of the photoreceptor differentiation sequence. The phenotype of a *spitz* null mutation is less severe than mutation in the *D-EGFR*, suggesting that like the mammalian receptor, other proteins may activate the D-EGFR. One candidate protein is gurken, another TGF-α like molecule, which is localized asymmetrically in the developing oocyte and may interact with the D-EGFR to determine dorso-ventral patterning of follicle cells.[59] However, gurken is unlikely to be important in the eye as it appears to only be required maternally and the eyes of gurken mutant flies are normal.

Spitz RNA is found across the eye disk, increasing at the morphogenetic furrow and continuing at a high level throughout the posterior developing region.[58] Both the D-EGFR and spitz are expressed more widely than their apparent range of activity, based on phenotypic studies of mutant alleles. This anomaly may explained by the activity of a number of genes which influence signaling by the D-EGFR. These genes include members of the *spitz* family and *argos*. These genes are of particular interest as they point to an additional complexity in the interaction of the D-EGFR and its ligand, which may have relevance to signaling by the EGFR in mammals. The *spitz* family is a group of genes which appear to act in a common pathway, based on the similarity of the phenotype of mutations in these genes (embryonic lethality associated with ventral ectodermal defects[60]). Members of the *spitz* family are *spitz*, *star*, *rhomboid* and *pointed*. The position at which *star*, *rhomboid*, *pointed* and *argos* operate in the pathway is only now being resolved. Argos inactivates the D-EGFR, whereas star and rhomboid appear to function upstream of the D-EGFR. Pointed is a nuclear protein required downstream in the pathway and is discussed in section VI.

Argos

argos (also known as *giant lens* and *strawberry*) was isolated independently by several groups.[61-63] Hypomorphic alleles of *argos* result in reduced viability and an extremely rough eye phenotype, whereas null alleles result in embryonic lethality. Argos is a secreted protein of 444 amino acids, with a single EGF domain. Ommatidia which are mutant for argos have additional photoreceptor, cone and pigment cells, indicating that argos functions to negatively regulate development of photoreceptor and accessory cells in the eye. Experiments in which mutant clones of varying size are generated show that argos can diffuse several cell diameters.[61] When large clones are generated, the majority of genetically mutant ommatidia adjacent to normal tissue are phenotypically normal, whereas most ommatidia in the center of the clones show the mutant phenotype.

In the wing loss, argos closely resembles the Ellipse mutation and can be suppressed by halving the dosage of the D-EGFR or of spitz. Conversely, overexpression of argos mimics the wing vein phenotype of hypomorphic *D-EGFR* alleles and this effect is enhanced by reducing the dosage of the D-EGFR or spitz and can be suppressed by overexpression of the D-EGFR.[54] Similar interactions are seen in the eye. These findings support the view that argos acts by negatively regulating the D-EGFR. This interaction appears to operate at the level of the receptor, as the addition of argos to an in vitro cell assay significantly reduces spitz-mediated autophosphorylation of the D-EGFR.[54] This suggests that argos either competes with spitz for binding to the D-EGFR or binds to another receptor which in turn negatively regulates D-EGFR activation. Further genetic analysis has been performed recently in an attempt to identify other proteins which interact with argos.[64] This analysis identified three genes, *bulge*, *soba* and *clown* whose loss can suppress the phenotype caused by overexpression of argos and therefore appear to be required for argos function. A dominant allele of *bulge* ($bulge^D$) has a similar phenotype to that cause by overexpression of argos; this is suppressed by soba and clown mutations.

Star and Rhomboid

A variety of experiments implicate rhomboid, a putative seven-transmembrane protein,[65] in regulation of signaling by the D-EGFR in the developing eye and in the wing vein.[57,66] Loss of rhomboid results in a similar phenotype in the ventral embryonic ectoderm to D-EGFR mutations. Clones of homozygous mutant rhomboid cells show a partially penetrant loss of the same cells that require spitz in the eye, R2/R5 and R8.[67] Overexpression of rhomboid in the eye results in a highly penetrant phenotype in which there is formation of additional photoreceptors.[67] Importantly, this phenotype can be suppressed by heterozygous null mutations in spitz, D-EGFR, Star, or other components downstream of the receptor.[57] Similarly, the phenotypic effects of ectopic expression of rhomboid in the wing are suppressed by a heterozygous null mutation in *D-EGFR*.[66]

Star encodes a protein of 598 amino acids with no sequence homology to other proteins, but which is likely to be a transmembrane protein.[68] Although it is required throughout development, its role is best characterized in the eye. Mosaic analysis shows that it is required in the same cells as those requiring spitz and rhomboid (R8, R2/R5).[69] Analysis of the expression pattern of an enhancer trap line indicates that Star expression commences immediately posterior to the morphogenetic furrow and that it is expressed in the R8, R2/R5 cells.[70] The phenotype of Star mutant embryos is enhanced by halving the dosage of D-EGFR, SOS or RAS,[57,70] and mosaic analysis has also shown that Star is required for wing vein formation.[70] Star appears to be downstream of rhomboid as mutation in Star suppresses the phenotype in the eye and wing associated with overexpression of rhomboid.[57,66]

Although nothing is presently known of the biochemical activities of rhomboid and Star proteins, these genetic analyses place Star and rhomboid upstream of the D-EGFR. It has been suggested that their function may be to limit the sites of D-EGFR and spitz interaction,[66] perhaps by regulating the availability of spitz for the D-EGFR. The restricted expression patterns of Star and rhomboid, compared with spitz and the D-EGFR, is consistent with this idea. The following experiment also supports the notion that rhomboid and Star may be required for the formation of mature spitz protein. Overexpression of the secreted, but not the precursor, form of spitz results in the inappropriate formation of ventral structures in transgenic embryos, consistent with ectopic activation of the D-EGFR signaling pathway.[49] This phenotype is not altered in embryos mutant for rhomboid or Star, indicating that, unlike the precursor protein, rhomboid and Star are not required upstream of secreted spitz in this assay.

IV. LINKING RAS WITH TYROSINE KINASE RECEPTORS

The previous section dealt with three receptors controlling separate cell fate decisions. A remarkable finding is that the signals downstream of these receptors, to a large extent, funnel into a common signaling pathway. This section describes the components of this pathway which lie between the D-EGFR, sevenless, torso and RAS.

RAS

Low-stringency cross hybridization with a v-Ha-RAS probe was used in the initial identification of three genes, *Dras 1-3*, with a high degree of homology to mammalian RAS proteins.[71] Dras-1 is 75% identical to human Ha-, K-, and N-RAS and appears to be the only functional homologue of these human proteins in *Drosophila*. Dras-2 is most closely related to R-RAS and its role in *Drosophila* is unknown at present. Revision of the initial Dras-3 sequence made it clear that Dras-3 is a homologue of human Rap and it therefore was renamed Rap1.[72] Rap1 appears to be antagonistic to the signal from sevenless.[72] Rho and Rac-family homologues also exist in *Drosophila* but will not be discussed here.[73]

An understanding of the involvement of Dras-1 in signaling by tyrosine kinase receptors in *Drosophila* emerged from the results of a series of elegant genetic experiments. These experiments began with a screen in which signaling by sevenless was made very sensitive to changes in the dosage of downstream signaling components (see also Fig. 1.3).[8] The basis of this screen was to raise transgenic flies which expressed a temperature sensitive allele of *sevenless*, at just below the restrictive temperature. Under these circumstances signaling by sevenless in wild type flies was sufficient for R7 cells to form, but the addition of heterozygous mutations in downstream components abolished signaling in this sensitized system. The mutations were termed enhancers of sevenless E(sev), as they enhanced the phenotype of the weak *sevenless* allele. The mutations obtained fell into seven complementation groups, of which four also suppressed the *Ellipse* allele of the *D-EGFR*, suggesting that these four genes were positively required for protein tyrosine kinase receptors in general. The latter genes have been cloned and are *Dras-1*, *Son of sevenless* (*SOS*),[74] *downstream of receptor kinases* (*Drk*)[75] and *corkscrew* (*csw*).[76]

The identification of Dras-1 as a positive regulator of sevenless signaling came at a time when similar data had been accumulating in other systems which linked tyrosine kinase receptors and RAS. For example, anti-RAS antibodies injected into PC12 cells had been shown to block NGF mediated neurite outgrowth, and dominant RAS alleles could mimic NGF action in the same system. The RAS gene *let60* was identified as also being downstream of the EGFR homologue, *let23*.[77] Expression of an constitutively active form of RAS in the eye results in the formation of supernumerary R7 cells, in a sevenless independent manner, providing further evidence that RAS activation is important in R7 cell differentiation.[78] The cells which adopt an R7 cell fate, the R7 precursor and the cone cells, are also sensitive to dominant alleles of *sevenless*. The R3/R4 photoreceptor cells, which appear to be refractory to sevenless signaling, are also unaffected by constitutively active RAS. This phenotype is not induced by an activated form of Dras-2, which is interesting considering that the homologous R-RAS protein has been shown to interact with Raf.[79] Clonal analysis demonstrates that Dras-1 is also involved in the determination of other photoreceptors, as patches of homozygous Dras-1 mutant ommatidia show greatly reduced formation of all photoreceptors.[8] As only the R7 photoreceptor responds to sevenless, the question remains as to the receptor which triggers RAS activation in the remaining photoreceptor cells. The D-EGFR is the most likely candidate for this, based on the phenotype of D-EGFR mutant clones in the eye (see section III). RAS activation has also been shown to be important for signaling by torso. Expression of constitutively active RAS can rescue the terminal phenotype in torso and corkscrew mutant embryos and expression of a dominant negative allele of RAS blocks terminal differentiation.[80]

These findings suggest that activation of RAS is sufficient to fully mimic signals from sevenless in the developing eye and from torso in the embryo. However, it is noteworthy that the activation of Raf, a downstream component of torso and sevenless, can be achieved in a partially RAS-independent manner and that the phenotype of RAS mutant embryos is less severe than that seen in torso or Raf mutants.[81] These observations are discussed further in the next section.

Son of Sevenless (SOS) and Downstream of Receptor Kinases (Drk)

The *SOS* gene was first identified as a dominant suppressor (*SOS*^{JC2}) of a weak *sevenless* allele.[82] (Its name follows the tradition that began with *bride of sevenless*!) It is not known whether the *SOS*^{JC2} mutation results in an enhanced SOS protein expression or activity. SOS was also discovered in the screen for enhancers of the *sevenless*^{ts} allele,[8] and subsequently in a screen for suppressors of a dominant *torso* allele.[83] A heterozygous mutation in SOS suppresses the phenotype in the eye of the dominant *Ellipse D-EGFR* allele.[8] Each of these studies therefore define SOS as being positively required for signaling downstream of the sevenless, D-EGFR or torso receptors. Its requirement downstream of the D-EGFR and torso probably accounts for the lethality seen in homozygous mutant SOS embryos. Cloning of the *Drosophila SOS*[8,84] and subsequently of mammalian homologues[85,86] showed SOS proteins to be large (1350-1500 amino acids), ubiquitously expressed, multidomain proteins.

The most telling feature of the SOS proteins, in terms of their likely function, is a domain of approximately 400 amino acids with a high degree of homology to the guanine nucleotide exchange domain present in a family of yeast proteins. These proteins have been characterized genetically and biochemically as activators of RAS proteins, by virtue of their ability to catalyze GDP-GTP exchange. Biochemical analyses have confirmed that the mammalian mSOS1 and huSOS1 are exchange factors for H-RAS and K-RAS.[87-89] Interestingly, huSOS1 shows marked enhancement of exchange activity for prenylated (or geranylgeranylated) H- and K-RAS compared with the unmodified RAS.[89]

Identification of an activator of RAS as being downstream of the D-EGFR and sevenless provided a further link in molecularly connecting RAS with tyrosine kinase receptors. However, as SOS lacks an

SH2 domain which might allow direct association with activated receptors, it appeared that additional proteins were required for this interaction. A confluence of biochemical studies defining SH3 binding sites,[90,91] and genetic experiments in *Drosophila* and *C. elegans*,[75,92,93] provided the next clue in this regard. The C-terminal regions of SOS proteins are rather poorly conserved between mammals and insects, with most of the homology clustered around proline rich peptides.[85] It has been shown in a number of laboratories that these peptides mediate the interaction of SOS with the adaptor protein Grb2, and its *Drosophila* counterpart Drk, and that this allows interaction of this complex with activated receptors.[75,86,87,92,94-96] Grb2/Drk encodes a single SH2 domain flanked by two SH3 domains. The SH3 domains of Grb2, particularly the N-terminal domain,[97] interact with the proline rich peptides of SOS. This complex is then recruited to activated receptors via the specific binding of the SH2 domain of Grb2 to phosphotyrosine residues. This model predicts that recruitment of SOS, which is normally cytosolic, to the membrane via the binding of Grb2 to activated receptors allows SOS to interact with membrane bound RAS. While this model is discussed in more detail in chapter 3, it is worth describing here a recent in vivo analysis of *Drosophila* SOS[98] which suggests that the actual situation may be more complex and may also involve the DH and PH domains of SOS.

The amino terminal region of SOS has two closely associated protein homology regions; the dbl-homology (DH) and pleckstrin homology domains (PH). The DH domain corresponds to a region in the *dbl* oncogene which has guanine nucleotide exchange activity for the RAS-related protein cdc42hs,[99,100] although such activity has not been described for the SOS DH domain. The PH domain was first recognized as a repeated motif in the PKC substrate protein pleckstrin and is present in a large number of proteins involved in signal transduction.[101] PH and DH domains are frequently found to be associated. The function of the PH domain is unclear, although it can mediate specific protein-protein interactions[102] or interact with membrane phospholipids.[103-105] The latter results are particularly interesting in view of the results of an in vivo structure-function analysis of *Drosophila* SOS.[98] In these experiments the activity of various mutant SOS proteins was assayed in developing flies by testing the ability of the mutant protein to contribute to R7 cell formation by the SOS^{JC2} allele. Although this assay system was somewhat artificial, it is noteworthy that they found (contrary to predictions of the above model) that deletion of the C-terminal Grb2 binding region of SOS did not affect SOS-mediated R7 cell formation. In contrast, deletion of the PH and DH domains abolished SOS function. This suggests that either Drk is dispensable in this assay or that C-terminal truncation of SOS renders Drk dispensable for SOS function. The importance of the DH and PH domains for the activation of RAS by the mammalian SOS1 protein has also been demonstrated recently in a cell overexpression system.[106] Consistent with these results is the finding that although tyrosine 2546 in the sevenless receptor has been shown to be necessary for Drk binding, this residue is dispensable for signaling by sevenless.[107] It is clear that Drk functions downstream of sevenless, the D-EGFR[75,92] and torso[81] and upstream of RAS. However Drk's role may not be to localize SOS to activated receptors but rather to influence the activity of SOS in other ways. Recent results in mammalian cells indicate that the C-terminal domain of SOS may negatively regulate SOS activity[107a,107b] and that the role of Grb2 is to relieve this effect. The idea that Drk may be doing something other than targeting SOS to the membrane is also supported by the observation that C-terminally truncated SOS protein shows a similar subcellular distribution to wild type protein in mammalian fibroblasts[107b] and membrane localization of SOS is not dependent on Drk in *Drosophila*.[98] However, it is noteworthy that the C-terminus is also a major site of ligand stimulated SOS phosphorylation in mammalian cells and this also appears to strongly influence SOS activity. Therefore it is difficult to separate the effects of Drk binding to the C-terminus of SOS from the effects of phosphorylation of this region when interpreting experiments involving C-terminally truncated SOS proteins.

GAP1

Whereas SOS proteins are important in activating RAS by promoting GDP-GTP exchange, GTPase activating proteins (GAPs) facilitate the conversion of RAS to an inactive, GDP-bound form by stimulating their intrinsic GTPase activity.[108] The *Drosophila GAP1* gene[109] was cloned from an enhancer trap screen for genes with specific patterns of expression in the eye imaginal disk, in a screen for mutations that could suppress the phenotype of a hypomorphic *sevenless*^{ts} allele (essentially the reverse of the screen described in ref. 74), in a screen for mutant flies with abnormal numbers of neural projections into the brain[110] and in a screen for enhancers of the *SOS*^{JC2} allele.[111] GAP1 is a protein of 1165 amino acids with a Ca^{2+}-dependent phospholipid binding domain and a PH domain with a high degree of homology to the Bruton's tyrosine kinase.[112] However, unlike mammalian GAP, it lacks an SH2 domain.

Flies which are homozygous for a null allele of *GAP1* form R7 cells in the absence of sevenless, and the cone cells transdetermine to supernumerary R7 cells. This suggests that there is a sufficient amount of GTP-bound RAS in the absence of sevenless to trigger R7 cell formation. It also indicates that GAP1 is simply a negative regulator of RAS in the *Drosophila* eye. This contrasts with experiments in mammalian cells which have suggested that GAPs may also be participants in downstream signaling,[113] although this view is less favored now.

GAP1 is expressed in a highly specific and restricted pattern, and flies which are mutant for the gene are fully viable.[109] In contrast Dras-1, the likely target of GAP1, is required generally in *Drosophila* and therefore it is very likely that there are other GAPs in *Drosophila.* The conclusion that GAP1 represents a highly tissue specific GAP in *Drosophila* is supported by the identification of mammalian GAP1 homologues, $GAP1^{m}$,[114] and GAPIII,[112] which also have restricted patterns of expression.

Corkscrew (csw)

Loss-of-function mutations in *csw* result in a partial loss of terminal structures. Csw is a nonreceptor class, tyrosine phosphatase of 841 amino acids with two amino terminal SH2 domains which is expressed throughout development in the embryo.[39] Csw is most similar to the human PTP1C protein in overall architecture and within the phosphatase domain, although csw contains a PTPase insert which is absent in PTP1C. Like PTP1C, csw also has two SH2 domains.

Csw functions downstream of torso, in that a homozygous loss-of-function mutation in *csw* suppresses the phenotype of a dominant *torso* allele in early embryos.[39] Therefore csw appears to be positively required for torso signaling and also for sevenless signaling, as it was also isolated from the screen for enhancers of a *sevenless*^{ts} allele.[76] How csw functions in relationship to torso and sevenless is unclear. The phenotypes of csw mutations are less severe than those of the other terminal class genes, and loss-of-function mutations in csw do not completely block torso signaling, as judged by expression of tailless, a downstream target of torso.[39] Injection of oncogenic RAS into embryos lacking csw rescues the embryonic cuticular defect in csw embryos, indicating that csw is upstream of RAS in the torso pathway.[80] In contrast, Simon[76] indicates that activated RAS cannot induce photoreceptor formation in cells which lack csw, placing it downstream of or parallel to RAS in the eye. A heterozygous mutation in *csw* suppresses the supernumerary R7 cell phenotype of a dominant *sevenless* allele but has little effect on the phenotype of a dominant *RAS* allele. These seemingly contradictory findings could be explained if csw acts in a pathway parallel to RAS that is essential for photoreceptor formation, but for which a 50% reduction in csw is not limiting for the formation of extra R7 cells by oncogenic RAS.

Shc and cbl

A large body of work has implicated the Shc and cbl proteins as important proteins downstream of many protein tyrosine kinase receptors in mammals (reviewed in ref. 115). Recently Shc was cloned in *Drosophila* by screening an expression library with an antiserum raised to the mammalian protein.[116] The overall structure of the mammalian and *Drosophila* proteins is very similar, with highest regions of homology in the N-terminal PTB and C-terminal SH2 domains and a lower degree of similarity in the central Gly/Pro-rich region. The dShc protein binds to an activated chimeric D-EGFR but not to sevenless. Interestingly, the major site of tyrosine phosphorylation on mammalian Shc, and to which Grb2 binds, is not conserved in dShc, and Drk was not found associated with phosphorylated dShc in vivo. Other phosphotyrosine sites, which may be important in mediating an interaction between Shc/dShc and other proteins, are conserved. The dShc gene localizes to the third chromosome at position 67B-C but no mutations which affect signaling by protein tyrosine kinases are known to map to that position.

The *cbl* proto-oncogene product has recently been recognized as a major target of phosphorylation downstream of a variety of mammalian tyrosine and nontyrosine kinase receptors, including the EGF receptor.[117-119] A possible functional role for cbl was provided by genetic studies in *C. elegans* with the cbl homologue, sli-1. It has been shown to be a negative regulator of EGFR (*let23*) signaling, acting between the receptor and RAS.[120]

The biochemical role of cbl is unclear at present. It forms a complex with activated receptors via the binding of Grb2 and other SH3 domain-containing proteins via its C-terminal proline rich region (reviewed in ref. 121). It has a centrally located RING finger domain and a basic amino terminal domain, both of which are highly conserved with sli-1 and with a related mammalian protein, cbl-B.[120] A C-terminal truncation which extends just past the RING finger domain results in a truncated protein which is transforming and which can bind the EGF receptor in a phosphotyrosine dependent manner. This protein has been postulated to act as a dominant negative for endogenous cbl (Thien and Langdon, submitted). The *Drosophila* protein has been cloned recently by low stringency cross hybridization (Hime, Pai Dhungat and Bowtell, unpublished data). The amino terminal and RING finger domains are highly conserved with the mammalian and *C. elegans* proteins, but surprisingly it lacks the C-terminal proline rich region and cannot associate with Drk. This provides another example of the difference in the way in which Drk and its mammalian counterpart, Grb2, interact with proteins in the RAS pathway. The extent to which this reflects a fundamental difference between insects and mammals in RAS signaling is unclear at present. However it is notable that D-cbl localizes to chromosome position 66C8-66D1 and like dShc, no known mutations in signaling by sevenless or the D-EGFR are known to map to this position.

V. A KINASE CASCADE DOWNSTREAM OF RAS

Analysis of the involvement of Raf, MEK and MAPK in signal transduction downstream of RAS in *Drosophila* has yielded results similar to those found in mammalian cells. As these findings have been reviewed extensively in chapters 5 and 6, this section will focus on additional information provided from *Drosophila.* The recent cloning of the kinase suppressor of RAS gene (KSR) is of particular interest in this regard.

RAF

The *Drosophila Raf* gene was cloned by low stringency cross hybridization with the mammalian gene[122] and subsequently shown to correspond to the terminal class mutant l(1)polehole.[37] Analysis of dominant and recessive alleles of *D-Raf* have shown it to be required downstream of torso for development of terminal structures, the D-EGFR[123] and sevenless.[124]

Expression of a dominant allele of *D-Raf* in the eye results in transdetermination of the cone cells into R7 cells in a sevenless independent manner.[124] Flies which are hemizygous for a weak allele of *Raf* show greatly reduced R7 cell formation and a reduction in the number of other photoreceptors, indicating that it is also required downstream of other tyrosine kinases in the eye.[125,126] Embryos for which there has been no maternal contribution of D-Raf, but which do have a paternal contribution that supplies zygotic expression, have a phenotype very similar to torso mutant embryos. Those with neither a maternal or paternal contribution have little cuticular development,[81,127,128] demonstrating that D-Raf is involved in the development of additional parts of the embryo.

Recent analysis of the role of D-Raf in formation of the terminal structures has provided a number of additional insights into the activation of the D-Raf. By making germline mosaic clones which give rise to homozygous mutant eggs, it has been possible to compare the phenotypes of embryos lacking maternal contributions of torso, Drk, SOS, Dras-1 and D-Raf genes.[81] Embryos lacking torso or D-Raf have a similarly extreme terminal phenotype and show greatly reduced expression of two target genes, *tailless* and *huckebein* (see section VI). In contrast, embryos which are homozygous for a complete deletion of Dras-1 show a less severe phenotype and have residual expression of tailless and hkb. Embryos which are homozygous for a D-Raf mutation (C110:replacement of Arg217 by Leu) which abolishes its interaction with Dras-1, show a less severe terminal phenotype than D-Raf null mutants.*[81] These findings suggest that D-Raf activation requires both Dras-1-dependent and -independent signals. Embryos which are mutant for the *SOS* gene also show a less severe phenotype than those lacking torso or D-Raf, but surprisingly this is less severe than the Dras-1 mutant animals. This cannot be explained by residual activity of the *SOS* allele as it represents a complete deletion for the gene and therefore implies the presence of additional guanine nucleotide exchange

**Arg217 of D-Raf corresponds to Arg89 of mammalian c-Raf-1 (see chapter 5).*

factor(s) which can partially activate Dras-1 in the absence of SOS. Drk mutant animals also show terminal defects but these are less extreme than both the Dras-1 and SOS mutant embryos. The nature of the Dras-1-dependent and -independent signals for D-Raf activation are unclear. In mammalian cells RAS appears to be important in the membrane localization of Raf[129,130] but association with activated RAS is insufficient to activate Raf (see chapters 5 and 6).

The activity of the *D-Raf^C110^* allele, which ablates Dras-1/D-Raf interaction, is just below that required for viability and this phenotype is fully penetrant. In a screen for suppressors of the lethality associated with this allele, a number of intragenic and extragenic suppressors were isolated.[125] The strongest of the intragenic suppressors (Su3) is a mutation (Phe290 to Ile) in the cysteine finger motif of the CR-1 region. Interestingly, this mutation does not restore Dras-1/D-Raf interaction in a yeast two hybrid system.[‡81] The authors speculate that this region may be involved in reception of the RAS-independent signal for D-Raf activation, as 14.3.3 and a few other proteins also interact with the CR1 domain.[131]

MEK

Screens for extragenic suppressors of the lethality associated with the *D-Raf^C110^* allele have also identified gain-of-function alleles of a *MEK* homologue (*suppressor of Raf-Dsor*[125,132]). These alleles are also capable of suppressing the phenotype of weak alleles of *D-EGFR* and *sevenless*.[132] Loss-of-function alleles of *Dsor* were isolated from a screen for suppressors of an oncogenic Dras-1 protein expressed in the eye (ref. 133 and see below). The involvement of Dsor in the D-EGFR, sevenless and torso pathways is supported by an analysis of transgenic flies carrying a temperature sensitive allele of *MEK* that was generated by site directed mutagenesis.[134] Analysis of the phenotype of animals grown at a restrictive temperature for various times shows that they have defects in dorso-ventral patterning and terminal structure formation in the embryo and in wing vein and R7 cell formation in the adult.

Recently the *Drosophila* gene *hemipterous* has been cloned[135] and found to encode a likely homologue of Jun kinase kinase (JNKK). This kinase appears to function in a separate pathway to Dsor, as a dominant allele of *MAPK* (*rl^Sevenmaker^*, see below), which can suppress the phenotype associated with loss of Dsor, cannot suppress the hemipterous phenotype. Therefore, separate MAPK-dependent signaling pathways occur in *Drosophila*, as has been defined in other species (see reviews in refs. 136, 137). This is also relevant to the discussion of nuclear proteins mediating photoreceptor development, as D-Jun is important in this setting (see section VI below).

‡ Phe290 of D-Raf corresponds to Phe163 of mammalian c-Raf-1.

MAPK

The *Drosophila* homologue of *MAPK* (*ERK-A*) was cloned by low-stringency cross hybridization[138,139] and alleles of the *rolled* locus were subsequently shown to correspond to this gene.[140] Loss-of-function mutations in rolled were also isolated from a screen for suppressors of the rough eye phenotype induced by the expression of a dominant *D-Raf* allele,[140] and many alleles have been isolated independently as suppressors of a dominant *Dras-1* allele.[133] A dominant allele ($rl^{Sevenmaker}$)[141] of *rolled* was isolated in a screen for dominant suppressors of boss, by testing mutagenized flies for functional R7 cells (UV phototaxis). The $rl^{Sevenmaker}$ allele is sufficient for the formation of R7 cells in sevenless, Dras-1 and D-Raf backgrounds and can suppress the lethality associated with a weak *D-Raf* allele ($D\text{-}Raf^{C110}$). Similar interactions are seen with torso and D-EGFR mutants. The $\mathrm{rl}^{\mathrm{Sevenmaker}}$ mutation represents a single amino acid substitution in the kinase domain (D334N). This does not render the kinase constitutively active but rather results in a reduced sensitivity to dual specificity phosphatase activity.[142]

Phosphatase 2A (PP2A), β-Geranylgeranyl Transferase (βGGT-1) and Kinase Suppressor of RAS (ksr)

Recently a screen was performed which sought enhancers and suppressors of the phenotype caused by expression of a constitutively active *RAS* transgene in the developing eye.[133,143] This allele induces an externally apparent rough eye due to the disordered development of underlying photoreceptors (formation of supernumerary R7 cells). Heterozygous mutations in genes which are positive or negative regulators of the RAS signal were predicted to suppress or enhance the rough eye phenotype, respectively, when their dosage was reduced by 50%. As the severity of the rough eye phenotype was proportional to the number of copies of the *RAS* allele, it was anticipated that other components would be uncovered if their dosage was halved. Importantly, mutations which are normally lethal when homozygous recessive could be uncovered in this screen and, because the phenotype could be scored readily on live flies, it was possible to screen 850,000 mutagenized flies in a period of about 12 months.

The majority of the enhancers obtained were homozygous viable, with no distinct eye phenotype. However, several enhancers were identified for which multiple alleles were obtained, including the transcription factor yan (see below) and a homologue of phosphatase 2A (PP2A).

Many of the suppressors fell into 31 lethal complementation groups. Within these groups alleles of *D-Raf*, *Dsor* and *rolled* were recognized, consistent with previous studies demonstrating the importance of these proteins in mediating signaling by Dras-1. Alleles were tested for their ability to suppress a dominant *D-Raf* allele, so as to identify proteins which are required between Dras-1 and D-Raf. As some of the genes in this class could be involved in posttranslational modification of Dras-1, this was investigated by testing the ability of the mutations to suppress a constitutively active myristylated *Dras-1* allele. One gene (*SR2-2*, renamed *βGGT-1*) failed to suppress this modified Dras-1 protein and was found to encode the β subunit of type I geranylgeranyl transferase. This finding is consistent with the prediction that the *Drosophila* Dras-1 protein is geranylgeranylated rather than farnesylated, as the CAAX box of Dras1 contains a leucine at position "X".

A second gene (*SR3-1*, renamed *KSR*) was isolated which failed to suppress a dominant *D-Raf* allele but suppressed both the myristylated and nonmyristylated activated *Dras-1* allele.[143] Cloning of the *KSR* gene showed that is was predicted to produce a protein of 966 amino acids with several distinctive features. Ksr has a C-terminal domain with the eleven subdomains found in all eukaryotic kinases and which is most closely related to the Raf member kinases. Amino terminal to the kinase domain are four domains (CA1-4) which are conserved between KSR isolated from a distantly related *Drosophila* species (*D. virilis*) and human and mouse homologues.

The kinase domain has distinctive substitutions in conserved residues in subdomains VIb, VIII and II, including a lysine to arginine substitution in the invariant lysine in subdomain II which has been shown to be essential in the phosphotransfer reaction. However, as one *KSR* allele was isolated in which there is a mutation within a conserved region of the kinase domain it is likely that the kinase domain is functional. Residues in domains VIb and VIII which are indicative of serine/threonine versus tyrosine substrate specificity were uninformative, as the *Drosophila* sequence is typical of tyrosine kinases and the mammalian sequence resembles serine/threonine kinases. Sequences in subdomain VIII which are important for the recognition of MEK as a substrate for Raf are not conserved, and no interaction could be detected between KSR and Dsor (D-MEK). The CA1 domain is conserved between species but is unrelated to other proteins in the database. A weak allele of *KSR* ($KSR^{S\text{-}548}$) represents a two amino acid substitution in this domain and therefore it has a yet undefined function. Domain CA2 is a short proline rich peptide which potentially could mediate binding to SH3 domain containing proteins. The CA3 and CA4 domain show significant homology with the CR1 and CR2 domains of Raf member proteins. However, the RAS binding site I present in the Raf CR1 domain is absent from KSR, and no interaction with Dras-1 was seen with bacterially expressed proteins or in a yeast two hybrid assay.

Ksr is widely expressed and strong loss-of-function alleles are homozygous lethal. Flies homozygous for the weak allele $KSR^{S\text{-}548}$ show roughened eyes with the loss of R7, R1 and R6 cells. However, mosaic analysis showed that no clones of strong loss-of-function alleles could be recovered in the eyes, and homozygous mutant larvae die at the pupal stages lacking imaginal disks, indicating that the gene is also required for cell proliferation and survival. Ksr appears to also be involved in signaling by torso, as embryos derived from germline clones which lacked KSR had a terminal phenotype closely resembling loss-of-function torso alleles and lacked posterior expression of *tailless* and *huckebein* target genes in this pathway. Activated D-Raf partially rescued the terminal phenotype of KSR mutant embryos.

The function of KSR is unclear. Therrien et al[143] propose three possible models (see Fig. 1.4).

VI. NUCLEAR EVENTS

The forgoing discussion describes a signaling pathway which is remarkably highly conserved and appears to direct an extraordinary array of developmental decisions. However, as activated RAS protein can direct R7 cell formation and development of terminal embryonic structures or affect wing vein formation in different settings, there must exist in each of these tissues molecular differences in the responding, undifferentiated cells. These differences are likely to be manifest, and the RAS signal specifically interpreted, in the nucleus to give the appropriate developmental outcome.

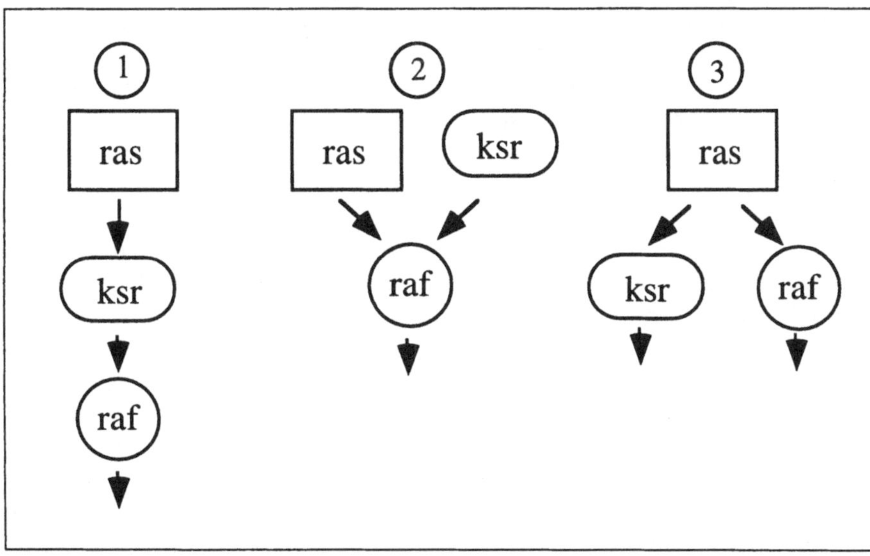

Fig. 1.4. Each model for KSR function is based on genetic data that strongly indicates that KSR lies either between RAS and Raf or in a parallel pathway to Raf, but not downstream of Raf. Although model 1 is the simplest explanation it does not explain the function of the previously demonstrated direct interaction RAS and Raf. Model 2 is attractive as KSR could represent the RAS-independent signal described by Hou et al.[81] The lack of interaction of KSR with known downstream targets in the sevenless and torso pathway (Dsor), the similarity of the KSR and Raf terminal phenotypes and the ability of activated D-Raf to suppress the KSR phenotype appear to make Model 3 less likely.

Nuclear Events in the Eye

Yan and pointed

The *yan* gene was isolated as a P-element insertion that produced a rough eye phenotype, associated with the formation of additional cone, pigment and photoreceptor cells, especially R7 cells.[144] Formation of R7 cells in a yan mutant is not dependent on the presence of sevenless or boss but does require another nuclear protein, sina (see below). Yan acts a negative regulator of differentiation in the eye and also in a wide range of neuronal and nonneuronal tissues.[145] The loss of yan is strongly synergistic with dominant *Dras-1* or loss-of-function *GAP1* alleles in the formation of extra R7 cells, implying that the function of yan is to inhibit Dras-1 mediated signaling.[144] *Yan* encodes a widely expressed protein of 732 amino acids which has an ETS domain, seen in a number of transcription factors, but most closely related to human elk1 and *Drosophila* E74.

Yan appears to be directly regulated by phosphorylation, probably by MAPK, and interacts with another ETS-domain protein, pointed, to control transcription of target genes. *Pointed* was initially identified as a gene which affects development of midline glial cells in the embryo.[146] It produces two alternatively spliced transcripts which give rise to two proteins, P1 and P2, which differ at their amino-terminal ends. Pointed is also required in the eye, as clonal analysis of strong *pointed* alleles or use of flies which are homozygous for a semi-lethal allele shows a reduction in photoreceptor development, especially of R7 cells.[147,148] Loss of pointed also suppresses the formation of supernumerary R7 cells associated with expression of oncogenic Dras-1[147] or dominant MAPK ($rl^{Sevenmaker}$).[148] Halving the dosage of pointed enhances the rough eye phenotype of weak *D-Raf* and *Dsor* alleles.[147] Pointed, like yan, therefore acts downstream of the RAS pathway in the determination of R7 cells but, in contrast, is positively required for their formation. The interplay between pointed and yan has become apparent through biochemical and genetic studies.

O'Neill et al[147] describe an in vitro cell transfection system in which expression plasmids for *yan*, *P1* and *P2* were transfected in various combinations and the expression of a CAT under control of a promoter with an ETS-consensus binding site driving was measured. The P1 protein strongly stimulated expression of the reporter plasmid, whereas the P2 protein did so more weakly, and only background activity was observed with yan alone. The activity of the P2 protein was greatly increased by cotransfection with dominant *Dras-1* or *MAPK* ($rl^{Sevenmaker}$) alleles, whereas the activity of P1 was unaffected. Cotransfection of *yan* with the *P1*-expressing plasmid inhibited P1 mediated activation of the reporter, indicating that yan directly inhibited the activity of pointed. The repression due to yan could be alleviated by additionally transfecting dominant *Dras-1* or $rl^{Sevenmaker}$. It seems likely that these effects are mediated by direct phosphorylation of both yan and pointed, as yan is phosphorylated in cells expressing dominant Dras-1 or $rl^{Sevenmaker}$, and mutation of a consensus MAPK site present in the P2 protein abolished the stimulatory effects of dominant Dras-1 and $rl^{Sevenmaker}$ for P2. Consistent with this, Brunner et al[148] demonstrate that yan and the P2 protein are substrates for the rolled protein in vitro.

These findings suggest a model where yan normally represses target genes required for R7 cell differentiation, and that activated MAPK both relieves this repression and activates the P2 protein. This model is supported by the finding that overexpression of P2 does not lead to formation of additional R7 cells in wild type flies but does so in flies which are heterozygous for yan.[148] More recently it has been shown that expression of a yan mutant protein which lacks all eight potential MAPK phosphorylation sites (yan^{ACT}) dominantly blocks neuronal differentiation in the eye and in embryonic tissues and the differentiation of many, but not all, nonneural tissues.[145] The protein expressed from this allele also shows enhanced stability in the eye disk, suggesting that phosphorylation of wild type yan increases its turnover. The wild type protein is translocated from the nucleus to the cytoplasm in cells expressing activated Dras-1, whereas yan^{ACT} remains in the nucleus. Interestingly, a dominant allele of *yan* (*Yan*^{S2382}) was isolated as a suppressor of the activated Dras-1 suppressor/enhancer screen described above. The protein is C-terminally truncated, removing two MAPK consensus sites and a PEST sequence, the latter being associated with protein degradation.[145]

The interplay between yan and pointed provides a mechanism for biochemically linking MAPK activation to transcription but leaves unresolved the issue of the physiological targets of these genes and the issue of specificity of RAS/MAPK activation of transcription. A recent series of papers involving the *Jun* and *phyllopod* genes provide some answers to these questions.

Jun

In mammalian cells Jun-family proteins heterodimerize with fos-family proteins to form the AP-1 family of transcription factors. Jun proteins are by regulated by phosphorylation by MAPK family proteins, primarily JNK (reviewed in refs. 136, 137). It appears that *Drosophila* has single Jun and fos proteins, known as D-Jun, jra or dm-Jun and D-fos or fra, respectively.[149-151] D-Jun is expressed transiently in the developing photoreceptors and cone cells in a sequence that parallels the differentiation sequence in these cells.[152] Using a dominant negative *D-Jun* allele (*Dm-Jun*^{bZIP}), consisting of the carboxy-terminal 107 amino acids of the protein and lacking the activation domain,[152] demonstrated a requirement for D-Jun in photoreceptor development. Overexpression of this protein in the eye causes a weakly penetrant loss of photoreceptors, including R1, R3, R4, R6 and R7. Loss of the

R7 photoreceptor was greatly enhanced in flies with a barely active *sevenless* allele, and the phenotype of dominant alleles of *sevenless*, *Dras-1*, *D-Raf* and *rolled* were suppressed by overexpression of Dm-Jun^{bZIP}. Collectively these findings indicated that D-Jun is important in RAS-mediated photoreceptor formation.

Further support for this conclusion has come from a recent analysis of the interaction of D-Jun with pointed and yan. The sites of phosphorylation of the human Jun protein have been mapped and conversion of these sites to phosphate-mimicking aspartic acid residues (JUN^{ASP}) results in a constitutively active protein.[153] Conversely, conversion of these residues to alanine results in a dominant negative protein (JUN^{ALA}). Expression of JUN^{ASP} in the eye induces the formation of supernumerary R7 cells, which are derived from the cone cells, in a sevenless independent manner.[154] Conversely, expression of JUN^{ALA} results in a rough eye phenotype associated with loss of R7 cells and some of the remaining photoreceptors. Importantly, these alleles interact with yan and pointed, both in an in vitro expression assay and in vivo. JUN^{ASP}-induced transformation of cone cells to R7 cells is enhanced in a yan heterozygous background and suppressed in a background where pointed activity is severely reduced. These findings are particularly relevant because they imply that the phenotype of the JUN^{ASP} allele requires pointed and remains susceptible to regulation by yan. This was confirmed in vitro by testing the ability of a JUN^{ASP} plasmid to cooperate with pointed in driving the expression of a AP1-ETS promoter/reporter construct.[154] Synergistic interaction was observed between JUN^{ASP} and pointed, and this could be partially suppressed by coexpression of yan. If D-Jun and pointed interact to drive expression of genes required for R7 cell specification, what are the targets of this interaction? The product of the *phyllopod* gene appears to be a very attractive candidate for a number of reasons.

Phyllopod

Phyllopod was cloned independently from screens for suppressors of dominant alleles of *D-Raf*[155] or *Dras-1*.[156] *Phyllopod* encodes a novel 400 amino acid nuclear protein with a central basic domain and C-terminal acidic domain, consistent with it being a transcription factor. Strong loss-of-function alleles of *phyllopod* are homozygous lethal, possibly because it is also required in the embryonic peripheral nervous system. Flies carrying a semiviable/null heterozygous allelic combination survive to the adult stage and show a loss of photoreceptors R1, R6 and R7, due to transformation of these cells into cone cells. Mosaic analysis using a null allele of *phyllopod* also shows loss of R1, R6 and R7 in mutant clones, and indicates that the gene is only required in the precursors of these cells for the formation of normal numbers of photoreceptors in each ommatidium.[156] Significantly, RNA in situ hybridization only detects phyllopod expression in the precursors of R1, R6 and R7 in wild type eyes. The supernumerary R7 cell phenotypes of gain-of-function alleles of *rolled* ($rl^{Sevenmaker}$) or *D-Raf* and loss of yan or GAP1 are suppressed by reduction of phyllopod. Ectopic expression of phyllopod in the cone cells results in their transdetermination to R7 cells. Importantly, ectopic expression of activated Dras-1 leads to ectopic phyllopod expression in the eye disk.

These results indicate that phyllopod is required, and its expression is induced, downstream of RAS activation and particularly, that it may be required in the interaction of yan and pointed. Phyllopod is the first gene which is expressed in the R7 precursor and not the cone cells (the R7 equivalence group) and whose expression appears to be dependent on the RAS signal. As Dras-1 is required in all photoreceptor cells, other genes must function to limit phyllopod expression to the R1, R6 and R7 cell precursors. The interaction of yan, pointed and Jun may be important in this regard, as loss of phyllopod also suppresses the supernumerary R7 cell phenotype induced by dominant Jun.[154] Given these findings, and that ectopic phyllopod expression induces R7 cell formation in cone cells, phyllopod fulfills many of the criteria for a gene which is a target of the RAS pathway and is then instructive for R7 cell formation in suitably primed cells. What the nature of that priming represents remains a central issue, as the R1 and R6 cells require phyllopod and yet respond differently to the cone and R7 cell precursors, as do other cells within the developing fly.

Sina

The *seven in absentia* gene (*sina*) was the first nuclear protein recognized as being specifically required for R7 cell formation,[157] and yet its function remains unclear. Flies which are homozygous for a null *sina* allele survive but the adults show reduced viability. There is complete loss of R7 photoreceptors, which become cone cells, and there is a variable loss of other photoreceptor cells. Mosaic analysis shows sina is only absolutely required in the R7 precursor. Sina is a 315 amino acid protein which localizes to both the cytoplasm and the nucleus.[157] The only clue to its function provided by its primary amino acid sequence is the presence of an N-terminal RING finger domain. The RING finger motif is a C3H-C4 domain which coordinates zinc ions to form an atypical zinc finger domain. It was originally described in the human *ring1* gene[158] and has since been identified in many proteins, including bmi-1, RAG1, c-cbl and BRCA1.[159,160] The function of most of these proteins is unclear, although some are implicated in transcriptional regulation, DNA repair, RNA translation regulation, and site-specific recombination. The role of the RING finger domain is itself unclear, although it may represent a novel mechanism of protein-protein or protein-nucleic acid interaction.

Significantly, functional sina protein must be present for the formation of normal numbers of R7 cells in the *Drosophila* eye or whenever supernumerary R7 cells are formed—for example, in response to dominant alleles of sevenless, RAS, Raf, rolled or D-Jun[23,126,141,148,154-156] or ectopic expression of proteins such as phyllopod or seven-up[155,156,161] or in the absence of yan.[144] However, unlike phyllopod or seven-up, overexpression of sina does not induce R7 cell formation and no dominant alleles of *sina* have been isolated to date. The phenotypic severity of weak alleles of *sina* are enhanced when flies also carry a mutation in rolled.[162] Collectively these observations place sina downstream, or in a parallel pathway to, all known proteins that contribute to R7 cell formation, but the biochemical basis of this is unclear. Flies which carry homozygous null *sina* mutations are viable, whereas null mutations in many of the other genes involved in R7 cell specification such as *Dras-1*, *D-Raf*, *rolled* and *SOS* are lethal. This implies that sina function is limited to a subset of the signaling pathways in which these other proteins operate. The mouse genome has three functional genes which are remarkably highly related *sina* homologues: *Siah-1A*, *Siah-1B* and *Siah-2*[163] and these may be involved in RAS mediated signaling in mammals.

Genes involved in specification of R1-R6 and R8:
seven-up*, *pipsqueak* and *rough

Much of this section of the review has focused on the control of R7 cell differentiation, as substantially less is known about differentiation of the remaining R1-R6 and R8 photoreceptors. Because of the development sequence in which these cells are formed (see section I), R2/R5, R3/R4 and R1/R6 can be considered as pairs. Based on their phenotype, expression of certain genes and their position in the mature ommatidium, R1-R6 are referred to collectively as the outer photoreceptors. Only a few genes are known to be specifically important for these cells: *rough*, *seven-up* and *pipsqueak*.

Rough encodes a homeobox protein[31] which is required in the R2/R5 photoreceptors for correct development of the adjacent R3 and R4 cells. In the absence of rough the R2/R5 cells default to become R3/R4 cells[164] and the subsequent sequence of photoreceptor development breaks down. Ectopic expression of rough in the R7 precursor causes it to adopt an R1-R6 fate, but surprisingly this requires the presence of sevenless and boss.[165,166] The signaling pathway controlling rough action is unknown.

The *seven-up* gene encodes a member of the 'orphan' steroid-receptor family. Recessive mutations in seven-up are embryonic lethal and its role in eye development first came to attention because of the pattern of expression of an seven-up enhancer trap in the developing eye disk.[167] Mosaic analysis has demonstrated that it is required in R1/R6 and R3/R4. Its expression in the wild type eye disk is limited to these cells, and in clones where it is absent there is transformation of the R1/R6 and R3/R4 into R7 cells. Misexpression of seven-up in the eye disk alters cell fate, but the outcome is dependent on the cell involved. For example, expression in R8 (the first photoreceptor to differentiate normally) leads to loss of R8 cells and secondary effects on R7 cell development.[168,169] Ectopic expression in R2/R5 mimics a rough mutation in that there is an absence of many of the R1/R6 and some R3/R4 cells, presumably due to the loss of cues from R2/R5. These effects underscore one of the difficulties in studying the control of R1-R6 development. That is, by perturbing the development of photoreceptors which differentiate early in the sequence, such as R8 and R2/R5, secondary developmental defects can occur which may be unrelated to the mutation. Ectopic expression of seven-up in the R7 cell and cone cells causes roughening of the exterior of the eye due to transdetermination of the cone cells into R7 cells and, at a low frequency, R7 cells to outer photoreceptor cells.[161,168,169] Several genes have been identified in a screen for suppressors of this phenotype, including loss-of-function alleles of *SOS*, *Dras-1* and *D-EGFR*, indicating that seven-up requires an intact RAS pathway to induce the formation of R7 cells.[161,169] The identification of the D-EGFR as a suppressor of this phenotype is particularly noteworthy, as it was not identified in other screens based on suppression of supernumerary R7 cells. Interestingly, if activated Dras-1 is coexpressed with seven-up the cone cells adopt an outer photoreceptor cell fate.[161] This implies that the D-EGFR normally provides a low level of RAS activity in the cone cell, which can be coopted in the presence of ectopically expressed seven-up to trigger R7 cell development. High levels of RAS activity in the cone cells may bring them to a state similar to the precursor of R1/R6 and R3/R4, where seven-up expression is required for outer photoreceptor fate. Whether the RAS pathway directly leads to the modification of seven-up or is simply required to make cells competent to respond to seven-up is unclear. Mutations in phyllopod suppress the cone cell to R7 cell phenotype induced by misexpression of seven-up,[161] indicating that phyllopod could be a target of seven-up. Recently it has been shown that ***pipsqueak***, which encodes a series of nuclear protein isoforms, is strongly expressed in R3/R4 and requires seven-up for its expression.[170] Strong loss-of-function alleles of ***pipsqueak*** result in a reduction in the number of R3/R4 photoreceptors and are dominant suppressors of misexpressed seven-up.

Nuclear Events in Torso Signaling

Tailless and Huckebein

The tailless and huckebein proteins are the best characterized nuclear targets of the torso pathway. Embryos which are mutant for both tailless and huckebein have defects in the development of the posterior terminal structures which resemble loss-of-function torso alleles, and the combined loss of tailless and huckebein strongly suppresses the effects of a gain-of-function torso allele.[34,171,172]

Tailless encodes a nuclear protein with homology to the steroid receptor class of transcription factors.[173] Although no ligand for tailless has been identified, the potential ligand binding domain is conserved with the mammalian Tlx protein, which has a high degree of overall sequence homology, and comparable functional activity, with tailless.[174] *Huckebein* encodes a C2H2 zinc finger protein.[175] The proteins are expressed in partially overlapping patterns at the termini of the developing embryo.[173,175,176] Control of the expression of huckebein and tailless at the anterior end of the embryo is complex, with input from the bicoid and dorsal pathways,[175,177] and consequently has been difficult to dissect. In contrast, their posterior expression appears to be primarily determined by torso activity and is better understood. Here the expression of the two proteins is very similar, appearing as caps of apparently uniform expression extending about 12% (huckebein) to 15% (tailless) of the length of the egg.[176] Expression of both proteins at the posterior end is absent in torso loss-of-function mutants and becomes generalized over the developing embryo in torso gain-of-function mutants.[176,178]

Although activation of the torso pathway is clearly necessary for tailless and huckebein expression, the link between regulation of these genes and the distal part of the torso pathway is not yet well characterized. Deletion analysis of the tailless regulatory sequences have defined regions whose removal results in generalized expression of a reporter construct, resembling the expression pattern seen in torso gain-of-function mutants.[179] Two proteins which specifically bind to these regions have been identified, GAGA factor and NTF-1, which correspond to the *Trithorax-like* and *grainyhead* genes, respectively. These findings suggest a model in which tailless expression is normally repressed in the central parts of the embryo by the binding of NTF-1 and/or GAGA to the tailless promoter, and that tailless is only expressed when this repression is alleviated by activation of the torso pathway. In support of this model NTF-1, which is a member of the ets-family of proteins, is phosphorylated by the mammalian MAPK, ERK2. Regulation of this class of proteins by the MAPK has precedents in mammalian cells (reviewed in ref. 180) and, as described above, in *Drosophila* with pointed and yan. However, control of tailless expression by torso is likely to be more complex than this model would suggest, as the 'repression domain' of the tailless promoter to which NTF-1 and GAGA bind confers only weak repression of a normally ubiquitously expressed heterologous promoter.[179] Genetic analysis of tailless and huckebein expression in GAGA and NTF-1 mutants is clearly important, but is incomplete at this stage, as germline mutant GAGA embryos could not be obtained from heterozygous mothers. However, mutants in NTF-1 show a partial expansion of the tailless expression in the posterior region of the embryo, supporting the notion that tailless expression is repressed somewhat by NTF-1.[179]

Tailless and huckebein are in turn important in regulating the patterns of expression of a number of early developmental genes of the central gap class, as judged by the expansion of the domains of expression of kruppel, knirps and giant in the absence of tailless and/or huckebein.[34,172,176,178] Tailless binds in vitro to sites in the *kruppel* promoter which are also occupied by bicoid and to a lesser extent by knirps[181] and it is likely that tailless represses kruppel expression in vivo by preventing bicoid-mediated transactivation. Binding of tailless to the *knirps* promoter has also been demonstrated.[182] Deletion of this part of the *knirps* promoter results in the expression of a β-galactosidase marker extending posteriorly, consistent with the view that tailless represses activation of knirps in this part of the embryo.

Bicoid

Not only does torso affect the ability of bicoid to activate genes such as kruppel through its effects on tailless expression but it also acts through a direct effect on bicoid itself. Genes such as *hunchback* and *orthodenticle* posses high affinity binding sites for the bicoid protein. The ability of bicoid to activate an artificial promoter consisting of three high affinity bicoid binding sites is repressed in embryos bearing a torso gain-of-function allele.[183] This may be mediated by phosphorylation of bicoid, as the protein is phosphorylated in a torso-dependent manner and most of the torso-mediated repression of bicoid is lost in D-Raf mutant embryos. How phosphorylation affects bicoid activity is not known. However, bicoid protein which is fused to a heterologous activation domain (GCN4) is resistant to torso-mediated repression.[183]

Dorsal

The torso pathway appears to also be important in regulating the activity of dorsal. The signaling pathway downstream of the toll receptor is important in dorso-ventral axis determination. Dorsal is a nuclear protein in this pathway which is both an activator of mesodermal and neuroectodermal target genes and also a repressor of genes such as *decapentaplegic* and *zerknullt*. The repressor activity of dorsal depends on the binding of corepressor proteins to the zen and dpp promoters (see ref. 184 and references therein). Several lines of evidence suggest that the torso pathway blocks the ability of dorsal to repress zen and dpp expression. Embryos which are mutant for a loss-of-function torso allele lose zen and dpp expression at their termini, but retain normal expression elsewhere, and zen is ubiquitously expressed in embryos expressing a gain-of-function torso allele. These findings could reflect either positive regulation of a zen and dpp transactivator, or inhibition of a repressor of these genes, by the torso pathway. The latter appears to be the case, as torso mutant embryos retain terminal zen and dpp expression, provided dorsal is also absent.[184]

These findings highlight an important aspect of torso regulation of gene expression, namely that this pathway serves to both activate expression of genes such as *tailless* and *huckebein* and also to limit the activity of other genes involved in body plan specification such as *dorsal*, *bicoid*, *kruppel*, *knirps* and *giant*. Both positive and negative regulation of gene expression by torso is therefore important in determining correct regional specification of the embryo.

VII. CONCLUDING REMARKS

The pace of discovery in these experimental systems has quickened greatly in the last five years, culminating in the description of unbroken pathways from the cell surface to the nucleus. Particularly striking has been the finding that most of the components downstream of torso, sevenless and D-EGFR are common to all three receptors, that they are present in organisms as diverse as *C. elegans* and mice and that they define a characteristic "signaling cassette". Although there may be subtle differences in the activation of these components by different receptors, it appears certain that signals from the RAS signaling cassette are to be interpreted specifically only when combined with other information which primes cells for a specific response to that signal. For example, this is seen in the eye where cells of the R7 equivalence group respond differently to a RAS signal than do the other cells comprising the ommatidium, and cells in the eye as a whole respond differently from those at the termini of the embryo. This suggests that the specificity of cellular response to RAS depends heavily on the progressive determination that occurs during development of a given tissue. The molecular basis of these differences between cells remains a major challenge of this field. Some insight is provided by systems such as the pheromone response in *S. cerevisiae*, where cell-specific responses to a common signal in **a** or α type cells is determined by the availability of cell-specific MAT proteins (reviewed in ref. 180). In *Drosophila*, the competence of a cell to respond in a specific way can be dependent on whether it is already developmentally committed when a RAS signal is provided. Hence, cells can adopt a different cell fate when there is loss of input from genes such as *seven-up* (R1, R6, R3, R4 become R7 cells) or *rough* (some R1-R6 cells become R7 cells). Ultimately the various signals must be integrated and culminate in the expression of genes which finally direct specific developmental outcomes (e.g. R7 cell formation). *phyllopod* is a good candidate for such a gene, as it is expressed in the R7 cell and not the cone cells, it can direct R7 expression when ectopically expressed and, importantly, its expression is RAS dependent (see discussion in refs. 185). However, even here the requirement for phyllopod in the R1 and R6 cells is difficult to explain.

Despite this gap in our understanding of how specificity is generated in the RAS pathway, this question should be tractable to the approaches described here. It therefore seems likely that the powerful genetics of *Drosophila* should continue to provide results and reagents which complement a wealth of biochemical studies in mammals.

Acknowledgments

I would like to thank Kevin Moses and Gary Hime for their comments on the manuscript.

References

1. Tomlinson A, Ready DF. Neuronal differentiation in the *Drosophila* ommatidium. Dev Biol 1987; 120:366-376.
2. Tomlinson A. The molecular basis of pattern formation in the developing compound eye of *Drosophila*. [Review]. Seminars in Cell Biology 1990; 1:229-39.
3. Cagan R. Cell fate specification in the developing *Drosophila* retina. [Review]. Development Supplement 1993; 19-28.
4. Ready DF. A multifaceted approach to neural development. Trends Neurosci 1989; 12:102-110.
5. Sprenger F, Nusslein-Vollard C. The terminal system of axis determination in the *Drosophila* embryo. In: The Development of *Drosophila Melanogaster*. Bate, Martinez M, Aris A, eds. Cold Spring Harbor Labaratory Press, 1993.
6. Duffy JB, Perrimon N. The torso pathway in *Drosophila*—Lessons on receptor tyrosine kinase signaling and pattern formation. Dev Biol 1994; 166:380-395.
7. Xu T, Rubin GM. Analysis of genetic mosaics in developing and adult *Drosophila* tissues. Development 1993; 117:1223-37.

7a. Golic KG, Lindquist S. The FLP recombinase of yeast catalyzes site-specific recombination in the *Drosophila* genome. Cell 1989; 59:499-509.

8. Simon MA, Bowtell DDL, Dodson GS et al. Ras1 and a putative guanine nucleotide exchange factor perform crucial steps in signaling by the sevenless protein tyrosine kinase. Cell 1991; 67:701-716.
9. Benzer S. Behavioral mutants of *Drosophila* isolated by countercurrent distribution. Proc Natl Acad Sci USA 1967; 58:1112-1119.
10. Tomlinson A, Ready DF. Sevenless: A cell specific homeotic mutation of the *Drosophila* eye. Science 1986; 231:400-402.
11. Harris WA, Stark WS, Walker JA. Genetic dissection of the photoreceptor system in the compound eye of *Drosophila melanogaster*. J Physiol 1976; 256:415-439.
12. Tomlinson A, Ready DF. Cell fate in the *Drosophila* ommatidium. Dev Biol 1987; 123:264-275.
13. Campos-Ortega JA, Jurgens G, Hofbauer A. Cell clones and pattern formation: Studies on sevenless a mutant of *Drosophila* melanogaster. Willhelm Roux' Archive of Dev Biol 1979; 186:27-50.

14. Hafen E, Basler K, Edstroem J-K et al. *sevenless*, a cell-specific homeotic gene of *Drosophila*, encodes a putative transmembrane receptor with a tyrosine kinase domain. Science 1987; 236:55-63.
15. Tomlinson A, Bowtell DDL, Hafen E et al. Localization of the *sevenless* protein, a putative receptor for positional information, in the eye imaginal disc of *Drosophila*. Cell 1987; 51:143-150.
16. Banerjee U, Renfranz PJ, Pollock JA et al. Molecular characterization and expression of sevenless, a gene involved in neuronal pattern formation in the *Drosophila* eye. Cell 1987; 49:281-91.
17. Tomlinson A, Bowtell DD, Hafen E et al. Localization of the sevenless protein, a putative receptor for positional information, in the eye imaginal disc of *Drosophila*. Cell 1987; 51:143-50.
18. Bowtell DDL, Simon MA, Rubin GM. Nucleotide sequence and structure of the *sevenless* gene of *Drosophila melanogaster*. Genes Dev 1988; 2:620-634.
19. Basler K, Hafen E. Control of photoreceptor cell fate by the sevenless protein requires a functional tyrosine kinase domain. Cell 1988; 54:299-311.
20. Mullins MC, Rubin GM. Isolation of temperature-sensitive mutations of the tyrosine kinase receptor sevenless (sev) in *Drosophila* and their use in determining its time of action. Proc Natl Acad Sci USA 1991; 88:9387-91.
21. Basler K, Hafen E. Ubiquitous expression of sevenless: position-dependent specification of cell fate. Science 1989; 243:931-4.
22. Bowtell DD, Simon MA, Rubin GM. Ommatidia in the developing *Drosophila* eye require and can respond to sevenless for only a restricted period. Cell 1989; 56:931-6.
23. Basler K, Christen B, Hafen E. Ligand-independent activation of the sevenless receptor tyrosine kinase changes the fate of cells in the developing *Drosophila* eye. Cell 1991; 64:1069-81.
24. Greenwald I, Rubin GM. Making a difference: the role of cell-cell interactions in establishing separate identities for equivalent cells. Cell 1992; 68:271-281.
25. Reinke R, Zipursky SL. Cell-cell interaction in the *Drosophila* retina: the *bride of sevenless* gene is required in photoreceptor R8 for R7 cell development. Cell 1988; 55:321-330.
26. Hart AC, Kramer H, Van VD Jr et al. Induction of cell fate in the *Drosophila* retina: the bride of sevenless protein is predicted to contain a large extracellular domain and seven transmembrane segments. Genes Dev 1990; 4:1835-47.
27. Krämer H, Cagan R, Zipursky SL. Interaction of *bride of sevenless* membrane-bound ligand and the *sevenless* tyrosine-kinase receptor. Nature 1991; 352:207-212.
28. Cagan RL, Krämer H, Hart AC et al. The bride of sevenless and sevenless interaction: internalization of a transmembrane ligand. Cell 1992; 69:393-399.
29. Hart AC, Kramer H, Zipursky SL. Extracellular domain of the boss trans-

membrane ligand acts as an antagonist of the sev receptor. Nature 1993; 361:732-6.
30. Van Vactor DL, Cagan RL, Krämer H et al. Induction in the developing compound eye of *Drosophila*: multiple mechanisms restrict R7 induction to a single retinal precursor cell. Cell 1991; 67:1145-1155.
31. Tomlinson A, Kimmel BE, Rubin GM. rough, a *Drosophila* homeobox gene required in photoreceptors R2 and R5 for inductive interactions in the developing eye. Cell 1988; 55:771-84.
32. Cagan RL, Kramer H, Hart AC et al. The bride of sevenless and sevenless interaction: internalization of a transmembrane ligand. Cell 1992; 69:393-9.
33. Schupbach T, Wieschaus E. Maternal-effect mutations altering the anterior-posterior pattern of the *Drosophila* embryo. Roux's Arch Dev Biol 1986; 195:302-317.
34. Klingler M, Erdelyi M, Szabad J et al. Function of torso in determining the terminal anlagen of the *Drosophila* embryo. Nature 1988; 335:275-7.
35. Szabad J, Erdelyi M, Hoffmann G et al. Isolation and characterization of dominant female sterile mutations of *Drosophila* melanogaster. II. Mutations on the second chromosome. Genetics 1989; 122:823-35.
36. Casanova J, Struhl G. Localized surface activity of torso, a receptor tyrosine kinase, specifies terminal body pattern in *Drosophila*. Genes Dev 1989; 3:2025-38.
37. Ambrosio L, Mahowald AP, Perrimon N. Requirement of the *Drosophila* raf homologue for torso function. Nature 1989; 342:288-91.
38. Stevens LM, Frohnhofer HG, Klingler M et al. Localized requirement for torso-like expression in follicle cells for development of terminal anlagen of the *Drosophila* embryo. Nature 1990; 346:660-3.
39. Perkins LA, Larsen I, Perrimon N. corkscrew encodes a putative protein tyrosine phosphatase that functions to transduce the terminal signal from the receptor tyrosine kinase torso. Cell 1992; 70:225-36.
40. Sprenger F, Stevens LM, Nusslein-Volhard C. The *Drosophila* gene torso encodes a putative receptor tyrosine kinase. Nature 1989; 338:478-83.
41. Sprenger F, Trosclair MM, Morrison DK. Biochemical analysis of torso and D-raf during *Drosophila* embryogenesis: implications for terminal signal transduction. Mol Cell Biol 1993; 13:1163-72.
42. Martin JR, Raibaud A, Ollo R. Terminal pattern elements in *Drosophila* embryo induced by the torso-like protein. Nature 1994; 367:741-745.
43. Savantbhonsale S, Montell DJ. Torso-like encodes the localized determinant of *Drosophila* terminal pattern formation. Genes Dev 1993; 7:2548-2555.
44. Casanova J, Furriols M, Mccormick CA et al. Similarities between trunk and spatzle, putative extracellular ligands specifying body pattern in *Drosophila*. Genes Dev 1995; 9:2539-2544.
45. Schupbach T, Wieschaus E. Germline autonomy of maternal-effect mutations altering the embryonic body pattern of *Drosophila*. Dev Biol 1986; 113:443-8.

46. Casanova J, Struhl G. The torso receptor localizes as well as transduces the spatial singal specifying terminal body pattern in *Drosophila*. Nature 1993; 362:152-155.
47. Sprenger F, Nusslein-Volhard C. Torso receptor activity is regulated by a diffusible ligand produced at the extracellular terminal regions of the *Drosophila* egg. Cell 1992; 71:987-1001.
48. Livneh E, Glazer L, Segal D et al. The *Drosophila* EGF receptor gene homolog: conservation of both hormone binding and kinase domains. Cell 1985; 40:599-607.
49. Schweitzer R, Shaharabany M, Seger R et al. Secreted spitz triggers the der signaling pathway and is a limiting component in embryonic ventral ectoderm determination. Genes Dev 1995; 9: 1518-1529.
50. Clifford R, Schupbach T. Molecular analysis of the *Drosophila* egf receptor homolog reveals that several genetically defined classes of alleles cluster in subdomains of the receptor protein. Genetics 1994; 137:531-550.
51. Baker NE, Rubin GM. Effect on eye development of dominant mutations in *Drosophila* homologue of the EGF receptor. Nature 1989; 340:150-3.
52. Baker NE, Rubin GM. Ellipse mutations in the *Drosophila* homologue of the EGF receptor affect pattern formation, cell division, and cell death in eye imaginal disks. Dev Biol 1992; 150:381-96.
53. Diaz-Benjumea FJ, Hafen E. The sevenless signaling cassette mediates *Drosophila* EGF receptor function during epidermal development. Development 1994; 120:569-78.
54. Schweitzer R, Howes R, Smith R et al. Inhibition of *Drosophila* egf receptor activation by the secreted protein argos. Nature 1995; 376:699-702.
55. Sawamoto K, Okano H, Kobayakawa Y et al. The function of argos in regulating cell fate decisions during *Drosophila* eye and wing vein development. Dev Biol 1994; 164:267-276.
56. Rutledge BJ, Zhang K, Bier E et al. The *Drosophila* spitz gene encodes a putative EGF-like growth factor involved in dorso-ventral axis formation and neurogenesis. Genes Dev 1992; 6:1503-1517.
57. Freeman M. The spitz gene is required for photoreceptor determination in the *Drosophila* eye where it interacts with the egf receptor. Mech Dev 1994; 48:25-33.
58. Tio M, Ma CY, Moses K. Spitz, a *Drosophila* homolog of transforming growth factor-alpha, is required in the founding photoreceptor cells of the compound eye facets. Mech Dev 1994; 48:13-23.
59. Neuman-Silberberg FS, Schupbach T. The *Drosophila* dorso-ventral patterning gene gurken produces dorsally localized RNA and encodes a TGF-α-like protein. Cell 1993; 75:165-174.
60. Mayer U, Nusslein-Volhard C. A group of genes required for pattern formation in the ventral ectoderm of the *Drosophila* embryo. Genes Dev 1988; 2:1496-1511.
61. Freeman M, Klambt C, Goodman CS et al. The argos gene encodes a diffusible factor that regulates cell fate decisions in the *Drosophila* eye.

Cell 1992; 69:963-75.
62. Okano H, Hayashi S, Tanimura T et al. Regulation of *Drosophila* neural development by a putative secreted protein. Differentiation 1992; 52:1-11.
63. Kretzschmar D, Brunner A, Wiersdorff V et al. Giant lens, a gene involved in cell determination and axon guidance in the visual system of *Drosophila* melanogaster. EMBO J 1992; 11:2531-9.
64. Wemmer T, Klambt C. A genetic analysis of the *Drosophila* closely linked interacting genes bulge, argos and soba. Genetics 1995; 140:629-641.
65. Bier E, Jan LY, Jan YN. rhomboid, a gene required for dorsoventral axis establishment and peripheral nervous system development in *Drosophila* melanogaster [published erratum appears in Genes Dev 1990 Apr;4(4): 680-1]. Genes Dev 1990; 4:190-203.
66. Sturtevant MA, Roark M, Bier E. The *Drosophila*-rhomboid gene mediates the localized formation of wing veins and interacts genetically with components of the egf-r signaling pathway. Genes Dev 1993; 7:961-973.
67. Freeman M, Kimmel BE, Rubin GM. Identifying targets of the rough homeobox gene of *Drosophila*: evidence that rhomboid functions in eye development. Development 1992; 116:335-46.
68. Kolodkin AL, Pickup AT, Lin DM et al. Characterization of star and its interactions with sevenless and egf receptor during photoreceptor cell development in *Drosophila*. Development 1994; 120:1731-1745.
69. Heberlein U, Rubin GM. *Star* is required in a subset of photoreceptor cells in the developing *Drosophila* retina and displays dosage sensititve interactions with *rough*. Dev Biol 1991; 144:353-361.
70. Heberlein U, Hariharan IK, Rubin GM. Star is required for neuronal differentiation in the *Drosophila* retina and displays dosage-sensitive interactions with *Ras1*. Dev Biol 1993; 160:51-63.
71. Neuman-Silberberg FS, Schejter E, Hoffmann FM et al. The *Drosophila* ras oncogenes: structure and nucleotide sequence. Cell 1984; 37:1027-33.
72. Hariharan IK, Carthew RW, Rubin GM. The *Drosophila* Roughened mutation: activation of a *rap* homolog disrupts eye development and interferes with cell determination. Cell 1991; 67:717-722.
73. Hariharan IK, Hu KQ, Asha H et al. Characterization of rho gtpase family homologues in *Drosophila* melanogaster—overexpressing rho1 in retinal cells causes a late developmental defect. EMBO J 1995; 14:292-302.
74. Simon MA, Bowtell DD, Dodson GS et al. Ras1 and a putative guanine nucleotide exchange factor perform crucial steps in signaling by the sevenless protein tyrosine kinase. Cell 1991; 67:701-16.
75. Simon MA, Dodson GS, Rubin GM. An SH3-SH2-SH3 protein is required for p21Ras1 activation and binds to sevenless and SOS proteins in vitro. Cell 1993; 73:169-77.
76. Simon MA. Signal transduction during the development of the R7 photoreceptor. Dev Biol 1994; 166:431-442.
77. Aroian RV, Koga M, Mendel JE et al. The let-23 gene necessary for Caenorhabditis elegans vulval induction encodes a tyrosine kinase of the EGF receptor subfamily [see comments]. Nature 1990; 348:693-9.

78. Fortini ME, Simon MA, Rubin GM. Signaling by the sevenless protein tyrosine kinase is mimicked by Ras1 activation [see comments]. Nature 1992; 355:559-61.
79. Spaargaren M, Martin GA, McCormick F et al. The Ras-related protein R-ras interacts directly with Raf-1 in a GTP-dependent manner. Biochem J 1994; 300:303-7.
80. Lu X, Chou TB, Williams NG et al. Control of cell fate determination by p21ras/Ras1, an essential component of torso signaling in *Drosophila*. Genes Dev 1993; 7:621-32.
81. Hou XS, Chou TB, Melnick MB et al. The torso receptor tyrosine kinase can activate Raf in a Ras-independent pathway. Cell 1995; 81:63-71.
82. Rogge RD, Karlovich CA, Banerjee U. Genetic dissection of a neurodevelopmental pathway: *Son of sevenless* functions downstream of the *sevenless* and EGF receptor tyrosine kinases. Cell 1991; 64:39-48.
83. Doyle HJ, Bishop JM. Torso, a receptor tyrosine kinase required for embryonic pattern formation, shares substrates with the sevenless and EGF-R pathways in *Drosophila*. Genes Dev 1993; 7:633-46.
84. Bonfini L, Karlovich CA, Dasgupta C et al. The Son of sevenless gene product: a putative activator of Ras. Science 1992; 255:603-6.
85. Bowtell D, Fu P, Simon M et al. Identification of murine homologues of the *Drosophila* son of sevenless gene: potential activators of ras. Proc Natl Acad Sci USA 1992; 89:6511-5.
86. Chardin P, Camonis JH, Gale NW et al. Human Sos1: a guanine nucleotide exchange factor for Ras that binds to GRB2. Science 1993; 260:1338-43.
87. Buday L, Downward J. Epidermal growth factor regulates p21ras through the formation of a complex of receptor, Grb2 adapter protein, and SOS nucleotide exchange factor. Cell 1993; 73:611-20.
88. Liu BX, Wei W, Broek D. The catalytic domain of the mouse sos1 gene product activates Ras proteins in vivo and in vitro. Oncogene 1993; 8:3081-4.
89. Porfiri E, Evans T, Chardin P et al. Prenylation of Ras proteins is required for efficient hSOS1-promoted guanine nucleotide exchange [published erratum appears in J Biol Chem 1994 Nov 11;269(45): 28522]. J Biol Chem 1994; 269:22672-7.
90. Cicchetti P, Mayer BJ, Thiel G et al. Identification of a protein that binds to the SH3 region of Abl and is similar to Bcr and GAP-rho. Science 1992; 257:803-6.
91. Ren R, Mayer BJ, Cicchetti P et al. Identification of a ten-amino acid proline-rich SH3 binding site. Science 1993; 259:1157-61.
92. Olivier JP, Raabe T, Henkemeyer M et al. A *Drosophila* SH2-SH3 adaptor protein implicated in coupling the sevenless tyrosine kinase to an activator of Ras guanine nucleotide exchange, SOS. Cell 1993; 73:179-91.
93. Clark SG, Stern MJ, Horvitz HR. *C. elegans* cell-signaling gene sem-5 encodes a protein with SH2 and SH3 domains. Nature 1992; 356:340-344.
94. Egan SE, Giddings BW, Brooks MW et al. Association of SOS Ras ex-

change protein with Grb2 is implicated in tyrosine kinase signal transduction and transformation [see comments]. Nature 1993; 363:45-51.
95. Rozakis-Adcock M, Fernley R, Wade J et al. The SH2 and SH3 domains of mammalian Grb2 couple the EGF receptor to the Ras activator mSos1 [see comments]. Nature 1993; 363:83-5.
96. Baltensperger K, Kozma LM, Cherniack AD et al. Binding of the ras activator son of sevenless to insulin receptor substrate-1 signaling complexes. Science 1993; 260:1950-1952.
97. Raabe T, Olivier JP, Dickson B et al. Biochemical and genetic analysis of the Drk SH2/SH3 adaptor protein of *Drosophila*. EMBO J 1995; 14:2509-18.
98. Karlovich CA, Bonfini L, McCollam L et al. In vivo functional analysis of the Ras exchange factor son of sevenless. Science 1995; 268:576-9.
99. Zheng Y, Hart MJ, Cerione RA. Guanine nucleotide exchange catalyzed by dbl oncogene product. Methods Enzymol 1995; 256:77-84.
100. Hart MJ, Eva A, Evans T et al. Catalysis of guanine nucleotide exchange on the CDC42Hs protein by the dbl oncogene product. Nature 1991; 354:311-4.
101. Musacchio A, Gibson T, Rice P et al. The PH domain: a common piece in the structural patchwork of signaling proteins. [Review]. Trends Biochem Sci 1993; 18:343-8.
102. Touhara K, Inglese J, Pitcher JA et al. Binding of G protein beta gamma-subunits to pleckstrin homology domains. J Biol Chem 1994; 269: 10217-20.
103. Garcia P, Gupta R, Shah S et al. The pleckstrin homology domain of phospholipase c-delta(1) binds with high affinity to phosphatidylinositol 4,5-bisphosphate in bilayer membranes. Biochemistry 1995; 34: 16228-16234.
104. Hyvonen M, Macias MJ, Nilges M et al. Structure of the binding site for inositol phosphates in a ph domain. EMBO J 1995; 14:4676-4685.
105. Lemmon MA, Ferguson KM, Obrien R et al. Specific and high-affinity binding of inositol phosphates to an isolated pleckstrin homology domain. Proc Natl Acad Sci USA 1995; 92:10472-10476.
106. McCollam L, Bonfini L, Karlovich CA et al. Functional roles for the pleckstrin and dbl homology regions in the ras exchange factor Son-of-sevenless. J Biochem 1995; 270:15954-15957.
107. Raabe T, Olivier JP, Dickson B et al. Biochemical and genetic analysis of the drk sh2/sh3 adaptor protein of *Drosophila*. EMBO J 1995; 14: 2509-2518.
107a. Aronheim A, Engelbert D, Li NX, Alalawi N, Schlessinger J, Karin M. Membrane targeting of the nucleotide exchange factor sos is sufficient for activating the ras signaling pathway. Cell 78:979-961.
107b. Wang W, Fischer EMC, Jia Q, Dunn JM, Porfiri E, Downward J, Egan SE. The Grb2 binding domain of mSos1 is not required for downstream signal transduction. Nature Genet 1995; 10: 294-300.
108. McCormick F. The GTPase superfamily. Introduction. [Review]. Ciba Foundation Symposium 1993; 176:1-5.

109. Gaul U, Mardon G, Rubin GM. A putative Ras GTPase activating protein acts as a negative regulator of signaling by the Sevenless receptor tyrosine kinase. Cell 1992; 68:1007-19.
110. Buckles GR, Smith ZD, Katz FN. mip causes hyperinnervation of a retinotopic map in *Drosophila* by excessive recruitment of R7 photoreceptor cells. Neuron 1992; 8:1015-29.
111. Rogge R, Cagan R, Majumdar A et al. Neuronal development in the *Drosophila* retina: the sextra gene defines an inhibitory component in the developmental pathway of R7 photoreceptor cells. Proc Natl Acad Sci USA 1992; 89:5271-5.
112. Baba H, Fuss B, Urano J et al. GapIII is a new brain-enriched member of the GTPase-activating protein family. J Neurosci Res 1995; 41:846-858.
113. Hall A. ras and GAP—who's controlling whom?. [Review]. Cell 1990; 61:921-3.
114. Maekawa M, Li SW, Iwamatsu A et al. A novel mammalian ras gtpase-activating protein which has phospholipid-binding and btk homology regions. Mol Cell Biol 1994; 14:6879-6885.
115. Margolis B, Skolnik EY. Activation of Ras by receptor tyrosine kinases. [Review]. J Amer Soc Nephrology 1994; 5:1288-99.
116. Lai K, Olivier JP, Gish GD et al. A *Drosophila* shc gene product is implicated in signaling by the der receptor tyrosine kinase. Mol Cell Biol 1995; 15:4810-4818.
117. Bowtell D, Langdon WY. The protein product of the c-cbl oncogene rapidly complexes with the egf receptor and is tyrosine phosphorylated following egf stimulation. Oncogene 1995; 11:1561-1567.
118. Meisner H, Czech MP. Coupling of the proto-oncogene product c-cbl to the epidermal growth factor receptor. J Biol Chem 1995; 270: 25332-25335.
119. Galisteo ML, Dikic I, Batzer AG et al. Tyrosine phosphorylation of the c-cbl proto-oncogene protein product and association with epidermal growth factor (egf) receptor upon egf stimulation. J Biol Chem 1995; 270:20242-20245.
120. Yoon CH, Lee JH, Jongeward GD et al. Similarity of sli-1, a regulator of vulval development in C. elegans, to the mammalian proto-oncogene c-cbl. Science 1995; 269:1102-1105.
121. Langdon WY. The cbl oncogene—a novel substrate of protein tyrosine kinases. Australian & New Zealand Journal of Medicine 1995; 25:859-864.
122. Mark GE, MacIntyre RJ, Digan ME et al. *Drosophila* melanogaster homologs of the raf oncogene. Mol Cell Biol 1987; 7:2134-40.
123. Brand AH, Perrimon N. Raf acts downstream of the egf receptor to determine dorsoventral polarity during *Drosophila* oogenesis. Genes Dev 1994; 8:629-639.
124. Dickson B, Sprenger F, Morrison D et al. Raf functions downstream of Ras1 in the Sevenless signal transduction pathway [see comments]. Nature 1992; 360:600-3.

125. Lu X, Melnick MB, Hsu JC et al. Genetic and molecular analyses of mutations involved in *Drosophila* raf signal transduction. EMBO J 1994; 13:2592-9.
126. Dickson B, Sprenger F, Morrison D et al. Raf functions downstream of Ras1 in the sevenless signal transduction pathway. Nature 1992; 360:600-603.
127. Ambrosio L, Mahowald AP, Perrimon N. l(1)pole hole is required maternally for pattern formation in the terminal regions of the embryo. Development 1989; 106:145-58.
128. Melnick MB, Perkins LA, Lee M et al. Developmental and molecular characterization of mutations in the *Drosophila*-raf serine threonine protein kinase. Development 1993; 118:127-138.
129. Leevers SJ, Paterson HF, Marshall CJ. Requirement for Ras in Raf activation is overcome by targeting Raf to the plasma membrane. Nature 1994; 369:411-4.
130. Stokoe D, Macdonald SG, Cadwallader K et al. Activation of Raf as a result of recruitment to the plasma membrane [see comments] [published erratum appears in Science 1994 Dec 16;266(5192): 1792-3]. Science 1994; 264:1463-7.
131. Freed E, Symons M, Macdonald SG et al. Binding of 14-3-3 proteins to the protein kinase Raf and effects on its activation. Science 1994; 265:1713-6.
132. Tsuda L, Inoue YH, Yoo MA et al. A protein kinase similar to MAP kinase activator acts downstream of the raf kinase in *Drosophila*. Cell 1993; 72:407-14.
133. Chang HC, Karim FD, Oneill EM et al. Ras signal transduction pathway in *Drosophila* eye development. Cold Spring Harbor Symposia on Quantitative Biology 1994; 59:147-153.
134. Hsu JC, Perrimon N. A temperature-sensitive MEK mutation demonstrates the conservation of the signaling pathways activated by receptor tyrosine kinases. Genes Dev 1994; 8:2176-87.
135. Gilse B, Bourbon H, Noselli S. hemipterous encodes a novel *Drosophila* MAP kinase kinase, required for epithelial sheet movement. Cell 1995; 83:451-461.
136. Karin M. The regulation of AP-1 activity by mitogen-activated protein kinases. J Biol Chem 1995; 270:16483-16486.
137. Karin M, Hunter T. Transcriptional control by protein phosphorylation—Signal transduction from the cell surface to the nucleus. Curr Biol 1995; 5:747-757.
138. Biggs WH III, Zipursky SL. Primary structure, expression, and signal-dependent tyrosine phosphorylation of a *Drosophila* homolog of extracellular signal-regulated kinase. Proc Natl Acad Sci USA 1992; 89:6295-6299.
139. Biggs WH, Zipursky SL. Primary structure, expression, and signal-dependent tyrosine phosphorylation of a *Drosophila* homolog of extracellular signal-related kinase (vol 89, pg 6295, 1992). Proc Natl Acad Sci USA 1993; 90:6377.

140. Biggs Wr, Zavitz KH, Dickson B et al. The *Drosophila* rolled locus encodes a MAP kinase required in the sevenless signal transduction pathway. EMBO J 1994; 13:1628-35.
141. Brunner D, Oellers N, Szabad J et al. A gain-of-function mutation in *Drosophila* MAP kinase activates multiple receptor tyrosine kinase signaling pathways. Cell 1994; 76:875-88.
142. Bott CM, Thorneycroft SG, Marshall CJ. The sevenmaker gain-of-function mutation in p42 map kinase leads to enhanced signaling and reduced sensitivity to dual specificity phosphatase action. Febs Letters 1994; 352:201-205.
143. Therrien M, Chang HC, Solomon NM et al. Ksr, a novel protein kinase required for ras signal transduction. Cell 1995; 83:879-888.
144. Lai ZC, Rubin GM. Negative control of photoreceptor development in *Drosophila* by the product of the yan gene, an ETS domain protein. Cell 1992; 70:609-20.
145. Rebay I, Rubin GM. Yan functions as a general inhibitor of differentiation and is negatively regulated by activation of the ras1/mapk pathway. Cell 1995; 81:857-866.
146. Klämbt C. The *Drosophila* gene pointed encodes two ETS-like proteins which are involved in the development of the midline glial cells. Development 1993; 117:163-176.
147. O'Neill EM, Rebay I, Tjian R et al. The activities of two Ets-related transcription factors required for *Drosophila* eye development are modulated by the Ras/MAPK pathway. Cell 1994; 78:137-47.
148. Brunner D, Ducker K, Oellers N et al. The ETS domain protein pointed-P2 is a target of MAP kinase in the sevenless signal transduction pathway. Nature 1994; 370:386-9.
149. Perkins KK, Dailey GM, Tjian R. Novel Jun- and Fos-related proteins in *Drosophila* are functionally homologous to enhancer factor AP-1. EMBO J 1988; 7:4265-73.
150. Perkins KK, Admon A, Patel N et al. The *Drosophila* Fos-related AP-1 protein is a developmentally regulated transcription factor. Genes Dev 1990; 4:822-34.
151. Zhang K, Chaillet JR, Perkins LA et al. *Drosophila* homolog of the mammalian jun oncogene is expressed during embryonic development and activates transcription in mammalian cells. Proc Natl Acad Sci USA 1990; 87:6281-5.
152. Bohmann D, Ellis MC, Staszewski LM et al. *Drosophila* jun mediates ras-dependent photoreceptor determination. Cell 1994; 78:973-986.
153. Papavassiliou AG, Treier M, Bohmann D. Intramolecular signal transduction in c-Jun. EMBO J 1995; 14:2014-9.
154. Treier M, Bohmann D, Mlodzik M. Jun cooperates with the ets domain protein pointed to induce photoreceptor r7 fate in the *Drosophila* eye. Cell 1995; 83:753-760.
155. Dickson BJ, Dominguez M, Vanderstraten A et al. Control of *Drosophila* photoreceptor cell fates by phyllopod, a novel nuclear protein acting down-

stream of the raf kinase. Cell 1995; 80:453-462.

156. Chang HC, Solomon NM, Wassarman DA et al. Phyllopod functions in the fate determination of a subset of photoreceptors in *Drosophila*. Cell 1995; 80:463-472.

157. Carthew RW, Rubin GM. seven in absentia, a gene required for specification of R7 cell fate in the *Drosophila* eye. Cell 1990; 63:561-577.

158. Lovering R, Hanson IM, Borden KL et al. Identification and preliminary characterization of a protein motif related to the zinc finger. Proc Natl Acad Sci USA 1993; 90:2112-6.

159. Miki H, Miura K, Matuoka K et al. Association of Ash/Grb-2 with dynamin through the Src homology 3 domain. J Biol Chem 1994; 269:5489-92.

160. Freemont PS. The RING finger: a novel protein sequence motif related to the zinc finger. Ann NY Acad Sci 1993; 684:174-192.

161. Begemann G, Michon AM, Vandervoorn L et al. The *Drosophila* orphan nuclear receptor seven-up requires the ras pathway for its function in photoreceptor determination. Development 1995; 121:225-235.

162. Carthew RW, Neufeld TP, Rubin GM. Identification of genes that interact with the sina gene in *Drosophila* eye development. Proc Natl Acad Sci USA 1994; 91:11689-93.

163. Della NG, Senior PV, Bowtell DDL. Isolation and characterisation of murine homologues of the *Drosophila* seven in absentia gene (sina). Development 1993; 117:1333-1343.

164. Heberlein U, Mlodzik M, Rubin GM. Cell-fate determination in the developing *Drosophila* eye: role of the rough gene. Development 1991; 112:703-12.

165. Basler K, Yen D, Tomlinson A et al. Reprogramming cell fate in the developing *Drosophila* retina: transformation of R7 cells by ectopic expression of rough. Genes Dev 1990; 4:728-39.

166. Kimmel BE, Heberlein U, Rubin GM. The homeo domain protein rough is expressed in a subset of cells in the developing *Drosophila* eye where it can specify photoreceptor cell subtype. Genes Dev 1990; 4:712-27.

167. Mlodzik M, Hiromi Y, Weber U et al. The *Drosophila* seven-up gene, a member of the steroid receptor gene superfamily, controls photoreceptor cell fates. Cell 1990; 60:211-224.

168. Hiromi Y, Mlodzik M, West SR et al. Ectopic expression of seven-up causes cell fate changes during ommatidial assembly. Development 1993; 118:1123-35.

169. Kramer S, West SR, Hiromi Y. Cell fate control in the *Drosophila* retina by the orphan receptor seven-up—its role in the decisions mediated by the ras signaling pathway. Development 1995; 121:1361-1372.

170. Weber U, Siegel V, Mlodzik M. Pipsqueak encodes a novel nuclear protein required downstream of seven-up for the development of photoreceptors r3 and r4. EMBO J 1995; 14:6247-6257.

171. Strecker TR, Halsell SR, Fisher WW et al. Reciprocal effects of hyper- and hypoactivity mutations in the *Drosophila* pattern gene torso. Science 1989; 243:1062-6.

172. Weigel D, Jurgens G, Klingler M et al. Two gap genes mediate maternal terminal pattern information in *Drosophila*. Science 1990; 248:495-8.
173. Pignoni F, Baldarelli RM, Steingrimsson E et al. The *Drosophila* gene tailless is expressed at the embryonic termini and is a member of the steroid receptor superfamily. Cell 1990; 62:151-63.
174. Yu RT, Mckeown M, Evans RM et al. Relationship between *Drosophila* gap gene tailless and a vertebrate nuclear receptor tlx. Nature 1994; 370:375-379.
175. Bronner G, Chulagraff Q, Doe CQ et al. Spl/egr-like zinc-finger protein required for endoderm specification and germ-layer formation in *Drosophila*. Nature 1994; 369:664-668.
176. Bronner G, Jackle H. Control and function of terminal gap gene activity in the posterior pole region of the *Drosophila* embryo. Mech Dev 1991; 35:205-11.
177. Pignoni F, Steingrimsson E, Lengyel JA. bicoid and the terminal system activate tailless expression in the early *Drosophila* embryo. Development 1994; 115:239-251.
178. Steingrimsson E, Pignoni F, Liaw GJ et al. Dual role of the *Drosophila* pattern gene tailless in embryonic termini. Science 1991; 254:418-21.
179. Liaw GJ, Rudolph KM, Huang JD et al. The torso response element binds gaga and ntf-1/elf-1, and regulates tailless by relief of repression. Genes Dev 1995; 9:3163-3176.
180. Hill CS, Treisman R. Transcriptional regulation by extracellular signals: mechanisms and specificity. [Review]. Cell 1995; 80: 199-211.
181. Hoch M, Gerwin N, Taubert H et al. Competition for overlapping sites in the regulatory region of the *Drosophila* gene Kruppel. Science 1992; 256:94-7.
182. Pankratz MJ, Busch M, Hoch M et al. Spatial control of the gap gene knirps in the *Drosophila* embryo by posterior morphogen system. Science 1992; 255:986-9.
183. Ronchi E, Treisman J, Dostatni N et al. Down-regulation of the *Drosophila* morphogen bicoid by the torso receptor-mediated signal transduction cascade. Cell 1993; 74:347-355.
184. Rusch J, Levine M. Regulation of the dorsal morphogen by the Toll and torso signaling pathways: a receptor tyrosine kinase selectively masks transcriptional repression. Genes Dev 1994; 8:1247-57.
185. Dickson B. Nuclear factors in sevenless signaling. Trends Genet 1995; 11:106-111.

CHAPTER 2

RAS-Mediated Signal Transduction in *C. elegans*

Min Han and Meera Sundaram

I. INTRODUCTION

The RAS-mediated signal transduction pathway, which is highly conserved among eukaryotes, plays key roles in multiple cellular and developmental processes including cell proliferation, differentiation and migration. In recent years, a combination of biochemical and genetic studies has allow scientists to identify many key players involved in RAS-mediated signal transduction and has begun to elucidate the mechanisms of activation and regulation of this pathway (see other chapters in this book). From receptor tyrosine kinases (RTKs) to RAS to MAP kinases, the main signal transduction pathway has been well established. The connections between this pathway and several other major signal transduction pathways have also begun to be revealed.

In this chapter, we review the recent studies on the function of a RAS pathway that controls vulval cell differentiation in the nematode *Caenorhabditis elegans* (*C. elegans*). Vulval induction in *C. elegans* provides us a unique opportunity to analyze the roles of RAS and other factors involved in RTK/RAS/MAP kinase-mediated signal transduction. Sensitive genetic screens have allowed us to identify not only many key factors such as *let-23 EGFR* and *let-60 RAS* acting in the main backbone of the signaling pathway, but also many positive and negative modifiers of this pathway. The knowledge accumulated from studies using the system has clearly established a specific function of this commonly used signaling pathway in controlling developmental

Regulation of the RAS Signaling Network, edited by Hiroshi Maruta and Antony W. Burgess.

pattern formation, and has also contributed significantly to the general understanding of the mechanisms of receptor tyrosine kinase/RAS-mediated signal transduction.

Vulval Differentiation in *C. elegans*

The vulva, an epithelial tube connecting the uterus to the outer environment, is the organ that hermaphrodites use to mate with males and to lay their eggs. However, both of these functions are "dispensable" in the laboratory: eggs can be fertilized and hatch inside the hermaphrodites. Therefore, animals with defective vulvae can continue to propagate, allowing easy genetic manipulation. Vulval cell lineages are also relatively simple and can be scored decisively at the individual cell level.

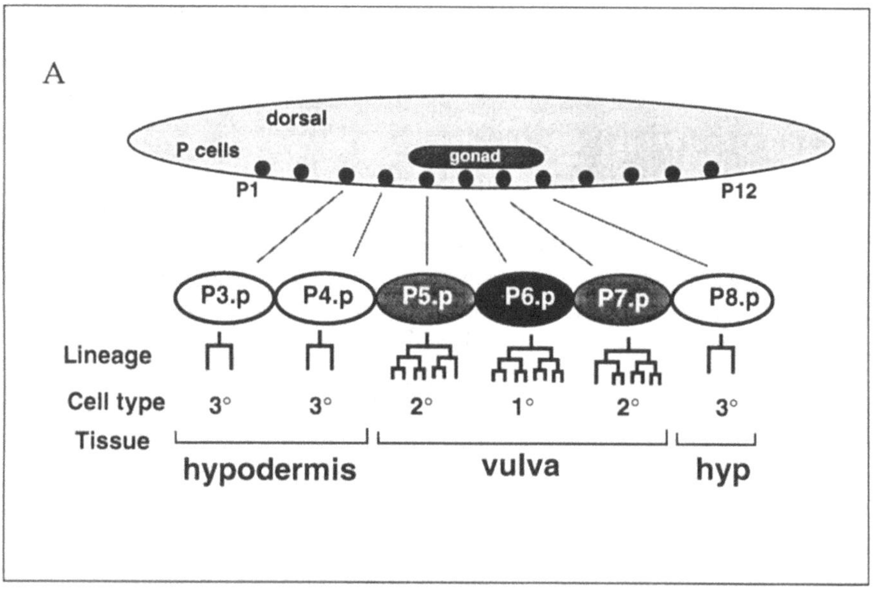

Fig. 2.1. Vulval cell differentiation is controlled by multiple cell-cell signaling events.

Fig. 2.1A. Schematic illustration of vulval precursor cells and their cell fates. Twelve hypodermal P cells migrate from sub-lateral sides of the animal to the ventral cord during the mid-L1 larval stage and divide once soon after the migration.[1,70] Vulval precursor cells (VPCs, P3-8.p) are the posterior daughters of six of these twelve P cells. During a period from the middle through the end of the L3 larval stage, the VPCs start to divide differentially. P3.p, P4.p, and P8.p divide only once and their progeny fuse with the large hypodermal syncytium hyp7 (3° cell fate). P5.p, P6.p, and P7.p divide further to generate vulval cells. The difference between 1° and 2° cell lineages is in the last round of cell divisions: P6.p (1°) produces eight progeny, while P5.p and P7.p (2°) each produce seven progeny. The progeny of 1° and 2° cells undergo distinct morphogenetic programs, and together with the anchor cell in the gonad, they form a vulva.[1,70]

The vulva is formed by 22 descendants of the posterior daughters of three ventral ectodermal P cells (Fig. 2.1).[1] Prior to vulval differentiation, six Pn.p cells (P3.p-P8.p, also known as vulval precursor cells or VPCs) have similar developmental potential. Each of them can either adopt a vulval fate or become nonspecialized hypodermal cells. In wild type, three of the six VPCs are induced to differentiate into two vulval cell types, 1° and 2°; each undergoes three rounds of cell division to generate eight or seven vulval descendants. The remaining three VPCs become hypodermal cells (3°) that divide only once to generate daughters that fuse with the surrounding hypodermal syncytium. The 3°-3°-2°-1°-2°-3° pattern (Fig. 2.1) of VPC fates is invariable among wild-type hermaphrodites.

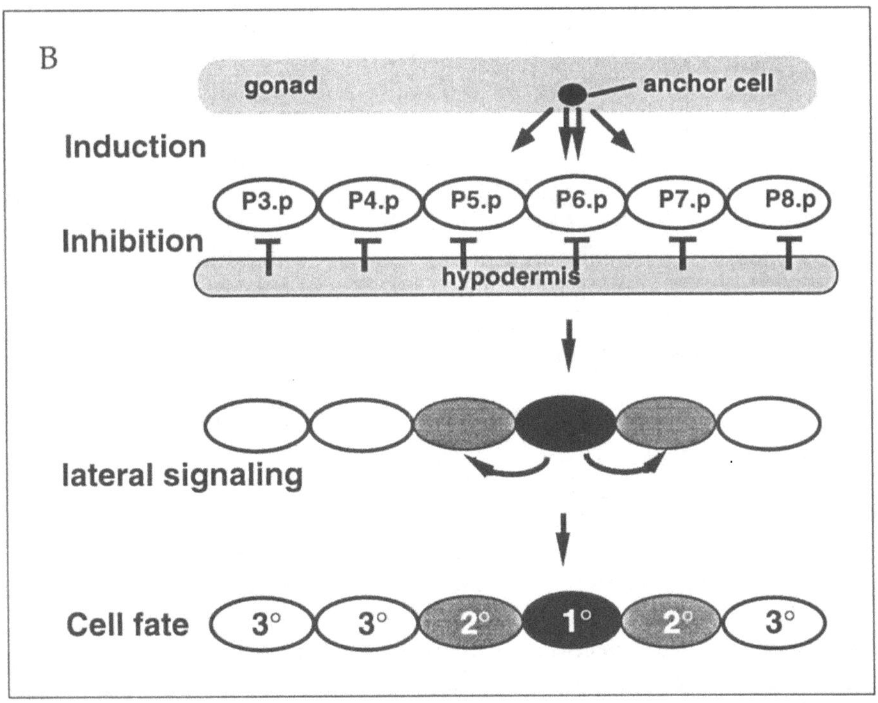

Fig. 2.1B. Current model for vulval cell fate specification by three separate signaling events. The six VPCs are initially equivalent and are capable of adopting either a vulval cell fate (1° or 2°) or a nonvulval fate (3°).[71,72] An inhibitory signal, possibly from the hypodermis, prevents each of the VPCs from adopting vulval cell fates.[6] However, an inductive signal from the anchor cell in the gonad overcomes this inhibition and promotes vulval fates in VPCs located near the anchor cell.[73] The 2°-1°-2° cell pattern for the three induced cells is likely generated by the combination of two different mechanisms. On one hand, the inductive signal may induce all three VPCs (P5-7.p) to adopt vulval cell fates: high levels of inductive signal promote the 1° vulval fate in P6.p (which is closest to the anchor cell), while lower levels of inductive signal promote the 2° fate in P5.p and P7.p.[4,72] On the other hand, high levels of inductive signal cause P6.p to produce a lateral signal that promotes the 2° vulval fate in its two neighboring VPCs.[44-46,74]

Multiple Signals Specify Vulval Cell Fates

Previous studies suggest that the precise pattern of vulval cell differentiation for the six VPCs is specified by the combination of three different intercellular signals (Fig. 2.1; for review see refs. 2, 3). (i) An inductive signal from a specialized cell (the anchor cell) in the gonad induces three of the six VPCs to adopt vulval cell fates. (ii) A lateral signal from a 1° VPC promotes the 2° fate and inhibits the 1° fate in neighboring VPCs. (iii) An inhibitory signal from the surrounding hypodermis may prevent VPCs from adopting vulval fates in the absence of inductive or lateral signaling.

According to the current model, the inductive signal formally acts to release the VPCs from the repression by the inhibitory signal: in the absence of the inductive signal, the vulval cell fates for the six VPCs are repressed by the inhibitory signal (all assume the 3° fate). In absence of the inhibitory signal, the VPCs all adopt vulval cell fates (1° and 2°) regardless of the presence of the inductive signal. The inductive signal is likely to be diffusible and spatially graded so that only VPCs relatively close to the anchor cell are induced to adopt vulval cell fates.[4] Proximity to the anchor cell also dictates which VPC will adopt the 1° cell type. The inductive signal from the anchor cell acts through an RTK-RAS-MAPK pathway that will be discussed extensively below. The existence of the hypodermal inhibitory signal was proposed based on the study of the *lin-15* gene products (also further discussed below).[5,6] The lateral signal acts through the *lin-12* gene product, a Notch-like receptor protein.[7]

II. GENETIC APPROACHES USED TO STUDY VULVAL SIGNAL TRANSDUCTION

Genetic studies on the vulval signaling pathway have not only allowed us to elucidate the roles of common signaling molecules in development, but also demonstrated the value of using a specific developmental genetic system to identify and characterize these signaling molecules. Mutations that result in mis-specification of VPC fates have defined many genes necessary for normal vulval cell differentiation. In these mutants, the cell fates of the six VPCs can be altered in opposite directions: they may all adopt vulval cell fates, generating extra vulval tissue (the "Multivulva" phenotype), or they may all become hypodermal cells leading to a "Vulvaless" phenotype (Fig. 2.2A).[8,9] The Vulvaless phenotype can be caused by loss of activity in the inductive signal or in the signal transduction pathway. The Multivulva phenotype may be caused either by loss of activity in negative regulators of the signal transduction pathway or by hyperactivity of positive factors in the pathway. So far, with no exception, every one of the molecules involved in the vulval signaling pathway (Fig. 2.3) was first identified by isolating genetic mutants.

Direct Screens for Vulvaless or Multivulva Mutants

Several key genes functioning in vulval differentiation were genetically identified by directly screening for mutations that cause defects in vulval cell lineages (Fig. 2.2B).[10-13] These initial screens were guided by specific morphological, structural, or functional defects associated with the mutants. For example, the *lin-1* gene was first identified by three "Multivulva" mutants that displayed extra ventral protrusions, while the *lin-3* gene was first identified by a mutant that was egg-laying defective.[10] These early screens based on the Vulvaless or Multivulva mutant phenotypes identified more than 10 genes that are involved in the signaling pathways that specify vulval cell fates. These genes include *lin-2, lin-3, lin-7, lin-10, lin-25,* and *let-23, lin-1, lin-8, lin-9, lin-13, lin-15, let-60(gf),* and *lin-12.* However, mutations in many key genes in the signaling pathway (including loss-of-function mutations in *sem-5, let-60, lin-45, mek-2/let-537* and *sur-1/mpk-1*) were not identified by these screens. The number of genomes in the early direct screens might not have been large enough to saturate for mutations. However, it became apparent later that all of the genes in the main signaling pathway (Fig. 2.3) play roles in multiple events during *C. elegans* development; null mutations in these genes often cause larval lethality as indicated by the names of several genes in the pathway (*let-23, let-60, let-341* and *let-537*). For many of these genes, mutations that cause specific defects in vulval development are rare and thus may not be recovered by direct screening for Vulvaless mutants.

Suppressor Screens

In the past few years, several laboratories have carried out extensive genetic screens for suppressors of existing Vulvaless or Multivulva mutants (Fig. 2.2C,D). These suppressor screens have been extremely effective in identifying genes involved in inductive signaling. For example, loss-of-function mutations in the *lin-15* gene disrupt the proposed hypodermal inhibitory signaling and cause all six VPCs to generate vulval tissue (Multivulva). Such a phenotype can be reverted (suppressed) by additional mutations that reduce the activity of genes acting downstream of *lin-15.* Screens for suppressors of *lin-15* mutations were carried out in two laboratories and resulted in identification of alleles of *let-23 EGFR, sem-5, let-341, let-60 RAS* and *lin-45 Raf.*[14-19] However, such screens did not identify mutations in genes acting downstream of Raf. The Multivulva phenotype associated with the *lin-15* mutations used in the screen may be too strong to be suppressed by subtle changes in the pathway.

More recently, screens for suppressors of the Multivulva phenotype caused by the activated *let-60 RAS* allele G13E have identified more than *120* mutations in at least 10 different genes, including *lin-45 Raf, mek-2, sur-1/mpk-1, lin-25, sur-2, ksr-1* and several others now

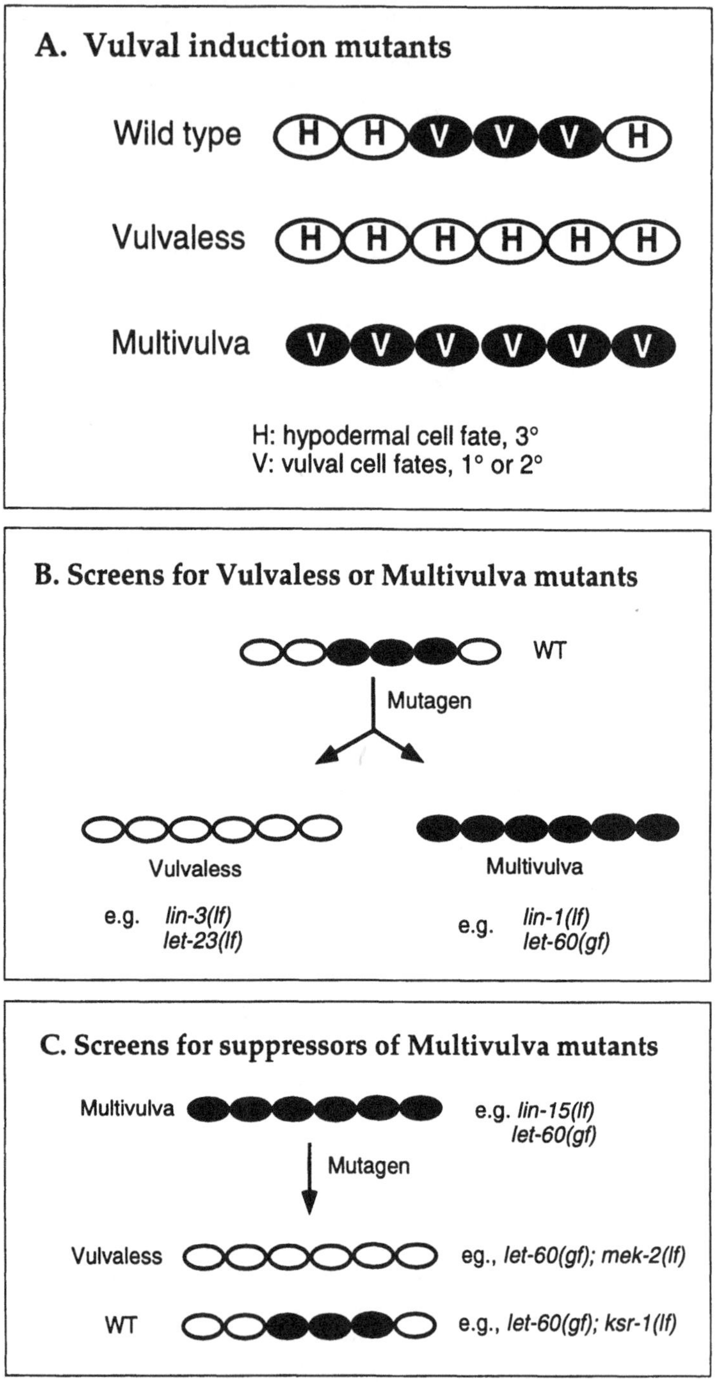

Fig. 2.2. Genetic approaches used to identify genes involved in vulval signaling.

A. Schematic illustration of the "Vulvaless" and "Multivulva" phenotypes. In Vulvaless mutants, zero or fewer than three VPCs adopt a vulval cell fate, resulting in animals that are egg-laying and mating defective. In Multivulva mutants, more than three VPCs adopt either the 1° or 2° vulval cell fate, resulting in animals with extra vulval tissue (pseudovulvae).

B. Direct screening for Vulvaless or Multivulva mutants to identify genes in the vulval signaling pathway.

C. Screening for mutations that suppress the Multivulva phenotype caused by mutations in known genes. These suppressor mutations may be dominant (may be recovered in the F1 generation of mutagenized parents) or recessive (recovered in the F2 generation). They may suppress the Multivulva phenotype and cause a Vulvaless phenotype, or they may suppress the Multivulva phenotype to wild type. These suppressor mutations are likely to define genes acting downstream of the gene whose mutant phenotype (Multivulva) is suppressed. For example, mutations that suppress the Multivulva phenotype of a lin-15 *mutation have been determined to be alleles of* let-23 EGFR, sem-5, let-341, let-60RAS *and* lin-45 Raf, *all of which act downstream of* lin-15.[14-19] *Mutations isolated as suppressors of activated* let-60 RAS *mutations define genes acting downstream of RAS such as* mek-2, sur-1/mpk-1, sur-2 *and* ksr-1.[20-25,27]

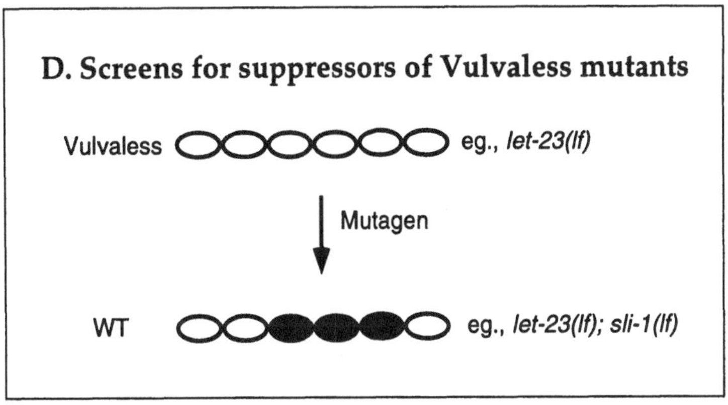

D. Screening for mutations that suppress the Vulvaless phenotype caused by mutations in known genes. Isolation of suppressors of the Vulvaless phenotype caused by mutations in the let-23 *EGFR gene has led to the identification of several negative regulators of LET-23-mediated signaling.*[28,29,55] *These suppressor mutations do not bypass the need for the LET-23 gene product, but significantly increase the signaling activity mediated by mutant LET-23 receptors. Such a strategy may also be used to identify additional negative regulators acting late in the signaling pathway.*

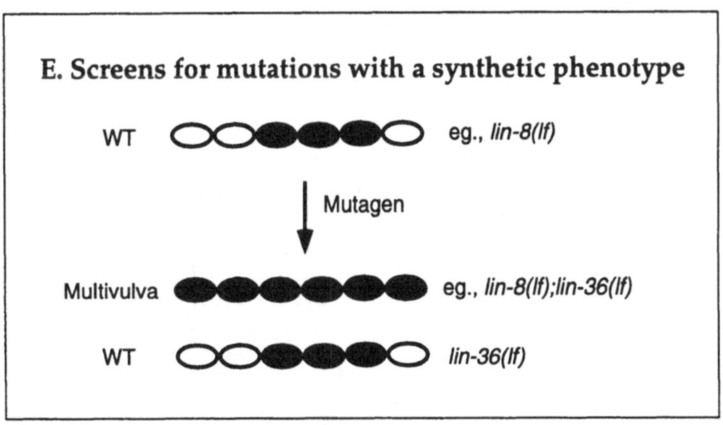

E. Screening for mutations that cause a synthetic mutant phenotype. If a gene function is redundant with that of another gene, a mutant phenotype may only be revealed by generating mutations in both genes. For example, a single loss-of-function mutation in a class A synthetic Multivulva gene such as lin-8 *does not perturb vulval differentiation. However, adding an additional loss-of-function mutation in a class B synthetic Multivulva gene such as* lin-36 *will generate a Multivulva phenotype. Such a strategy has been used to identify many new genes with redundant functions in vulval differentiation.*[5,55]

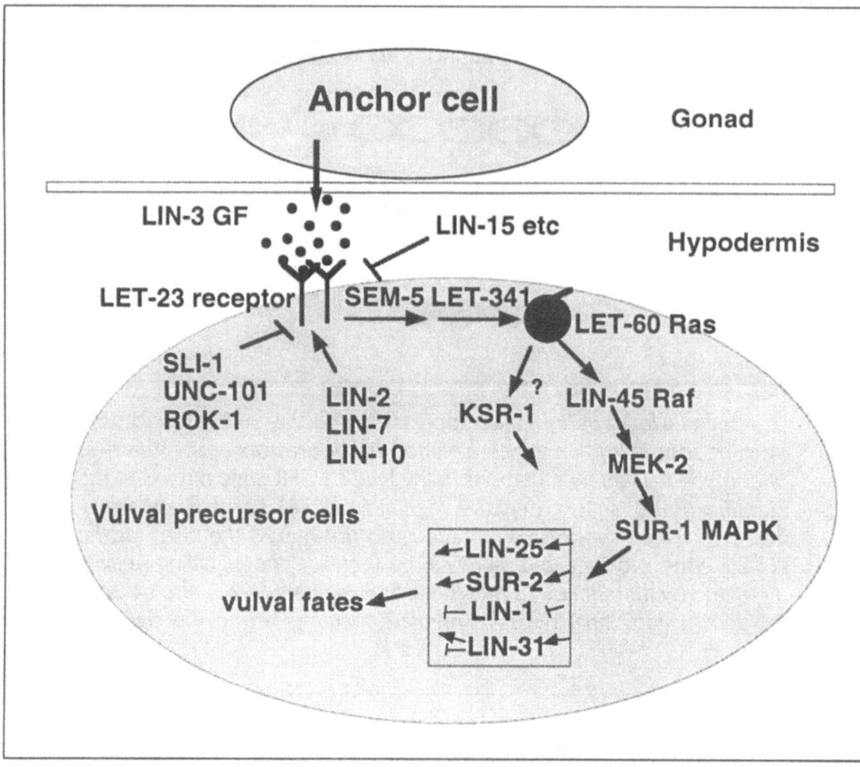

Fig. 2.3. Proposed genetic pathway that transduces the inductive signal to specify vulval cell fate. Arrows indicate positive regulation of one gene product by another. T bars represent negative regulation of one gene product by another. Neither arrows nor T bars necessarily represent a direct interaction. Biochemical studies in other systems have provided important information about the biochemical functions of many key factors in this signaling pathway. The lin-3 *gene encodes a growth factor that has been proposed to be the inductive signal from the anchor cell in the gonad.*[38] *The* let-23 *gene encodes a putative receptor tyrosine kinase (EGF receptor family) for the LIN-3 signal.*[42] *The* lin-15 *locus contains two genes (*lin-15A *and* lin-15B*) that may be components of the proposed inhibitory signal from the hypodermis.*[5,6,51,52] *The* sli-1, unc-101 *and* rok-1 *genes act negatively to regulate LET-23 RTK-mediated signaling processes, possibly by directly affecting the activity of LET-23.*[28,29,55,60] *The* lin-2, lin-7 *and* lin-10 *genes encode positive factors that may be required for proper function of LET-23 RTK.*[13,44,61,63] *SEM-5, an SH3-SH2-SH3 adaptor protein, binds to the phosphotyrosine of the activated LET-23 RTK via its SH2 domain, and presumably also to a guanine nucleotide exchange factor via its SH3 domains.*[17,49] *The* let-341 *gene, whose cloning has not been reported, is a candidate for encoding such an exchange factor.*[18] *The interaction between the receptor-bound SEM-5 and the exchange factor can then activate LET-60 RAS*[35] *by promoting the exchange from RAS-GDP to RAS-GTP. RAS-GTP would then bind to LIN-45 Raf and activate the LIN-45 Raf/MEK-2/SUR-1/MPK-1 kinase cascade.*[19-23,64] *The* ksr-1 *gene encodes a novel protein kinase that may act downstream of RAS but in parallel to Raf to positively regulate the signaling activity.*[24,27] *The* lin-1 *and* lin-31 *genes encode putative transcription factors that are likely to act downstream of SUR-1/MPK-1.*[67,68] *LIN-1 plays a negative role, but LIN-31 may play both positive and negative roles in the pathway. The* lin-25 *and* sur-2 *genes encode novel proteins that are positive factors acting downstream of SUR-1/MPK-1.*[25,26] *Their relationships with LIN-1 and LIN-31 are not understood.*

being characterized.[20-27] The success of such screens is partly due to the fact that the *let-60 RAS(G13E)* allele has an incompletely penetrant Multivulva phenotype which is highly sensitive to changes in the activity of genes acting downstream of RAS.

Negative regulators of the pathway may also be identified by screens for suppressors of the Vulvaless phenotype caused by mutations in genes in the pathway (Fig. 2.2D). For example, partial loss-of-function mutations in *let-23 RTK* cause a Vulvaless phenotype, and screens for suppressors of this phenotype have identified the *sli-1* and *unc-101* genes, two negative regulators of the pathway.[28,29]

ADVANTAGES OF SUPPRESSOR SCREENS

The screens for suppressors of a Multivulva or Vulvaless phenotype offer two distinct advantages over the early direct screens for a Vulvaless or Multivulva phenotype. The first advantage is the high specificity of the screens. The early hunts for egg-laying defective mutants recovered not only genes functioning in vulval cell differentiation, but also a large number of genes that function in other aspects of the egg-laying system or *C. elegans* development.[10,30] In contrast, by screening for mutations that revert a specific mutant phenotype, a specific effect of the new mutations is ensured. Furthermore, the suppressor hunts have also allowed us to concentrate on genes acting downstream of the gene whose mutant phenotype is reverted in the screen. For example, in the screens for suppressors of a *let-60 RAS(gf)* mutant, only genes acting downstream of RAS (e.g., *lin-45, mek-2, sur-1/mpk-1, sur-2,* and *lin-25*) were identified, while mutations in upstream genes such as *lin-2, lin-7, lin-10, lin-3, let-23* or *sem-5* were avoided.

The second distinct advantage is the high sensitivity of the screens. When proper strains are used, suppressor screens often allow the recovery of weak mutations that cause no phenotype or only a weak mutant phenotype by themselves. For example, the *let-60 RAS(G13E gf)* mutant background is much more sensitive than wild type to gene activity changes in downstream genes such as *lin-45 Raf, mek-2* and *sur-1/mpk-1.* "Silent" suppressor mutations that have no obvious phenotypes by themselves were isolated in all three genes mentioned above as well as in several new genes by screening for suppressors of the *let-60 RAS (G13E gf)* Multivulva phenotype.[20,22,27] (unpublished results in our laboratory; K. Kornfeld and R.H. Horvitz, personal communications). As discussed above, many genes involved in vulval signal transduction play multiple roles during *C. elegans* development and often have larval lethality or sterility as their null phenotype, so the ability to recover partial loss-of-function mutations that reveal a specific phenotype in vulval development is critical in identifying these genes.

Direct screening for Vulvaless or Multivulva mutants may also fail to identify genes that play subtle or redundant roles in signaling because these genes do not mutate to cause obvious vulval phenotypes.

The high sensitivity of suppressor screens may allow us to identify mutations in these genes. For example, severe mutations in the *ksr-1* gene do not cause a Vulvaless phenotype but almost completely suppress the Multivulva phenotype of the *let-60 RAS (G13E)* allele.[24,27]

For similar reasons, suppressor/enhancer screens also played huge roles in identifying factors acting in the *sev/RAS1* pathway that specifies the fate of the R7 photoreceptor in *Drosophila*.[31]

Genetic Interaction Studies

Genes involved in vulval induction have been ordered into a formal genetic pathway based on the results of epistasis analyses. For example, loss-of-function mutations in the *let-23 RTK* gene cause a Vulvaless phenotype due to a block in the signal transduction pathway, while loss-of-function mutations in the *lin-15* locus cause a Multivulva phenotype due to constitutive signal transduction resulting from the elimination of the proposed inhibitory signal from the hypodermis. The double mutants containing mutations in both genes display a Vulvaless phenotype like *let-23* mutants, indicating that *lin-15* acts upstream of *let-23*.[13,16] Similar experiments have also established, for example, that *let-60 RAS* acts downstream *let-23 EGFR* but upstream of *lin-45 Raf*.[14,15,19] However, the order of action among some genes such as *lin-45 Raf*, *mek-2* and *sur-1/mpk-1* has not been established by genetics because of a lack of gain-of-function mutations in these genes that cause a Multivulva phenotype. The functional relationships among these genes were proposed based on biochemical studies of similar gene products in mammalian cells and on the biochemical interactions between the worm proteins and mammalian proteins.[22]

Synergism analysis of mutations with similar phenotypes may also be informative about gene functions. If two genes' functions in vulval induction are redundant, an obvious mutant phenotype may not be seen in either single mutant but may be observed in double mutants containing mutations in both genes. As discussed below, synergistic analysis was used to study the function of *lin-15* and other synthetic Multivulva genes, the *sli-1* and *unc-101* genes, as well as the *ksr-1* gene (also see Fig. 2.2E).

One disadvantage of genetic interaction studies is that the results do not distinguish direct or indirect interaction between gene products. When gene A is said to act upstream of gene B, it could act several steps above gene B or it could be the direct regulator of gene B. Biochemical and molecular analysis of the gene product is necessary to understand the mechanism by which these signaling molecules function. The detailed biochemical studies of many conserved signaling molecules in mammalian cells have provided evidence for the biochemical roles of many genes functioning in the vulval signal transduction pathway.

III. GENETIC AND MOLECULAR ANALYSIS OF THE *LET-60 RAS* GENE

Genetic and molecular analyses of many different *let-60 RAS* mutations indicate that *let-60 RAS* acts as a genetic switch to control the choice between vulval and hypodermal cell fates of the six VPCs. The *let-60* gene was initially defined by recessive lethal mutations isolated during a general screen for such mutations.[32,33] Since then, more than 35 mutations in the *RAS* gene have been isolated in several laboratories, although only 25 of them have been reported (Table 2.1). The published mutations can be grouped into three different classes: recessive loss-of-function mutations (11), dominant negative mutations (9), and dominant gain-of-function (activated) mutations (5).

The reported eleven loss-of function mutations have been isolated as recessive lethal mutations,[32,33] recessive suppressors of the Multivulva phenotype of *lin-15* mutations,[15] or intragenic revertants of dominant negative mutations (Table 2.1).[14] Allele *sy127*, which results in a nonsense mutation at codon 123, appears to be the most severe mutation, presumably eliminating activity of the protein. The other loss-of-function mutations may define residues important for other aspects of *RAS* function. For example, Howe and Marshall[34] have made mutations in mammalian Ha-*RAS* based on some of the *let-60 (lf)* mutations, and found, for example, that mutations of amino acids 66 or 75 result in *RAS* protein that is unable to interact with an exchange factor, SCD25. *let-60 (lf)* mutations eliminate vulval induction, establishing the essential function of *let-60 RAS* in the vulval signaling pathway.[14,15]

Five activated (gain-of-function, or gf) *let-60 RAS* mutations were isolated and cause the identical G13 to E13 substitution (Table 2.1). Interestingly, these five mutations were isolated by five different investigators in five very different types of mutant screens: one by direct screening for Multivulva mutants, one by screening for suppressors of *let-23 RTK* mutations, one by screening for revertants of a *let-60 RAS* dominant negative allele, one by screening for suppressors of a *lin-10* mutation, and one by screening for suppressors of a mutation in the *let-341* gene.[12,14,15] Because most of these screens were done on a large scale, such a coincidence may suggest that other types of activated RAS mutations such as the common G12V mutation may not be recovered as viable alleles. The G13E mutation causes a semi-dominant Multivulva phenotype that is partially independent of the inductive signal. The Multivulva phenotype *of let-60 RAS(G13E)* is not completely penetrant (~160% vulval induction, comparing to 100% in wild type and 200% in completely penetrant Multivulva mutants) in homozygotes, and is highly sensitive to activity change downstream of RAS, making it an excellent strain for suppressor screens.

A Multivulva phenotype can also be generated by introducing multiple copies of the wild-type *let-60* gene into *C. elegans*.[27,35] High

Table 2.1. ***Genetic and molecular features of existing*** **let-60 RAS** ***mutations***

			Amino Acid				
Allele	**Class**[a]	**Phenotypes**[b]	**Codon**	**wild type**	**Mutant**	**Isolation**[c]	**Refs.**
s59	lf	recessive Let	136	Gly	Ser	lethality	14, 15, 32
s1155	lf	recessive Let	131, 132	Ser Glu	Cys Lys	lethality	15, 33
s1124	lf	recessive Let	37	Glu	Lys	lethality	14, 15, 33
sy101 sy163	lf	recessive Let	10, 110	Gly, Pro	Arg, Arg	Revertant of *sy101*	14, 36
sy101 sy127	lf	recessive Let	123	Arg	Stop	Revertant of *sy101*	14, 36
n1046 n1981	lf	recessive Let	13, 89	Gly, Ser	Glu, Phe	Revertant of *n1046*	15
n1876	lf	recessive Vul, Let	66	Ala	Thr	suppressor of *lin-15*	15
n2034	lf	recessive Vul, Let	66	Ala	Thr	suppressor of *lin-15*	15
n2035	lf	recessive Vul, Let	66	Ala	Val	suppressor of *lin-15*	15
n2022	lf	recessive Vul, Let	Splicing donor site in intron 2			suppressor of *lin-15*	15
n2021	lf	recessive Vul	75	Gly	Ser	suppressor of *lin-15*	15
n1046	gf	Multivulva	13	Gly	Glu	Mutivulva	12, 15
n1700	gf	Multivulva	13	Gly	Glu	suppressor of *lin-10*	15
n1849	gf	Multivulva	13	Gly	Glu	suppressor of *let-341*	15
sy130	gf	Multivulva	13	Gly	Glu	suppressor of *let-60(dn)*	14, 15
sy103	gf	Multivulva	13	Gly	Glu	suppressor of *let-23*	15
sy99	dn	dominant Vul, recessive Let	10	Gly	Arg	suppressor of *lin15*	14, 36
sy101	dn	dominant Vul, recessive Let	10	Gly	Arg	suppressor of *lin-15*	14, 36
n2301	dn	dominant Vul, recessive Let	15	Gly	Ser	suppressor of *lin-15*	15
n1531	dn	dominant Vul, recessive Let	15	Gly	Asp	suppressor of *lin-15*	15
sy94	dn	dominant Vul, recessive Let	16	Lys	Asn	suppressor of *lin-15*	14, 36
sy92	dn	dominant Vul, recessive Let	89	Ser	Phe	suppressor of *lin-15*	14, 36
sy95	dn	dominant Vul, recessive Let	89	Ser	Phe	suppressor of *lin-15*	14, 36
sy100	dn	dominant Vul, recessive Let	89	Ser	Phe	suppressor of *lin-15*	14, 36
sy93	dn	dominant Vul, recessive viable	119	Asp	Asn	suppressor of *lin-15*	14, 36

[a] lf stands for loss-of-function mutation (null or partial loss-of-function mutation). *sy127* is the only putative *let-60* null mutation.[35] gf stands for gain-of-function or hyperactive mutation, while dn stands for dominant negative mutation.
[b] Vul or vulvaless indicates a total or partial loss of vulval differentiation. Let or lethal indicates complete lethality. Only lethal and vulval phenotypes are listed. Many mutations cause other mutant phenotypes.
[c] The phenotype used to isolate each mutation is indicated.

copy number of either the wild-type or the activated form of the *let-60* gene can also result in larval lethality.

The nine dominant negative mutations (dn) were isolated as suppressors of the Multivulva phenotype of *lin-15* mutants.[14,15] All of these mutations cause a dominant Vulvaless phenotype and a male mating defect, and all but one of them cause recessive lethality at the first larval stage. These dominant negative mutations cause replacements of conserved residues that appear to be important for the RAS protein to bind to guanine nucleotides.[36] Genetic analysis has also shown that all LET-60(dn) proteins interfere with LET-60(+) activity, presumably by competing for an activator such as a guanine nucleotide exchange factor.

Interestingly, some dominant negative mutant proteins, while interfering with wild-type RAS protein, appear to retain partial constitutive activity to trigger vulval cell fate specification.[36] Transgenic animals containing several copies of the mutant genes display a Multivulva phenotype. For example, although *let-60(D119N)* has the strongest dominant negative effect among the nine dominant negative alleles, it is the only homozygous viable dominant negative allele and it causes a Multivulva phenotype when several copies of the mutant gene are introduced into animals. The constitutive nature of the residual activity may be due to a decrease in affinity of RAS for guanine nucleotides that leads to an increase in the exchange rate from RAS-GDP to RAS-GTP.[37] Therefore, some of the *let-60(dn)* alleles may result in two separable and abnormal biochemical functions of the protein: they may form a stable complex with an activator (likely exchange factor), preventing it from interacting with LET-60(+), and they may also retain a low level of activity that does not require the activator.

IV. GENES ACTING UPSTREAM OF RAS

THE *LIN-3* AND *LET-23* GENES: THE INDUCTIVE SIGNAL AND ITS RECEPTOR

Loss-of-function mutations in the *lin-3* and *let-23* genes cause a Vulvaless phenotype, establishing their essential function in vulval induction. Like *let-60 RAS*, null mutations in both *lin-3* and *let-23* are recessive lethal in the L1 larval stage, indicating their essential roles in early development.

The *lin-3* gene product, LIN-3, is proposed to be the inductive signal from the anchor cell based on several lines of experiments. (i) LIN-3 is structurally similar to growth factors of the EGF family which are membrane-spanning proteins with a single extracellular EGF domain.[38] (ii) Transgenic animals containing multiple copies of wild type *lin-3* can cause extra VPCs to assume vulval fates, and cell ablation experiments indicate that these transgenes act in the gonad to promote vulval development.[4,38] (iii) *lin-3:: lacZ* fusion transgenes that can induce vulval development are expressed specifically in the anchor cell at the time of vulval induction.[38] (iv) *lin-3* acts upstream of *let-23* EGFR since the Vulvaless phenotype of *lin-3* is suppressed by a *let-23* gain-of-function mutation and the Multivulva phenotype caused by *lin-3* transgenes is suppressed by *let-23* loss-of-function mutations.[38,39] Recently, Katz et al[4] have shown, by modulating the activity level of *lin-3* in transgenes, that different levels (concentrations) of LIN-3 protein promote distinct VPC cell fates: a high dose of LIN-3 induces the 1° vulval cell fate, while a lower dose of LIN-3 induces the 2° vulval cell fate. A high dose of LIN-3 can also override the lateral signaling that normally inhibits formation of adjacent 1° cell fates.

let-23 was originally defined by three larval lethal alleles[40,41] and a cold-sensitive, semi-lethal Vulvaless allele.[12] Additional *let-23* alleles have been isolated either by isolating suppressors of the Multivulva phenotype of *lin-15* mutations or by isolating mutations that fail to complement a *let-23* allele for vulval induction.[16,18] Genetic epistasis analysis indicates that *let-23* acts before *let-60 RAS* but after *lin-15* and *lin-3.*,[13,14,38] consistent with its role as the receptor of the anchor cell signal.

The overall architecture and sequence of LET-23 protein is similar to the EGF receptor subfamily of receptor tyrosine kinases.[42] Certain partial loss-of-function mutations of *let-23* preferentially disrupt vulval induction or cause other tissue specific mutant phenotypes.[16] DNA sequences of some of these *let-23* alleles suggest that the receptor's C-terminus can be divided into at least three domains that each contribute to receptor functions in different cell types.[43] These results suggest that the LET-23 EGFR protein may interact with multiple factors for tissue-specific functions. An activated *let-23* mutation (*sa62*) that causes an anchor cell-independent Multivulva phenotype has recently been isolated.[39] This mutation causes an amino acid substitution Cys to Tyr in the extracellular domain of the receptor and may result in ligand-independent activity. It has also been shown by genetic analysis that rare partial loss-of-function mutations in the *let-23* gene result in a hyperinduced phenotype (extra VPCs induced to vulval fate by the anchor cell signal).[16] Unlike the Multivulva phenotype caused by the *sa62* mutation, the hyperinduced phenotype is totally dependent on the inductive signal. These mutations might reduce a negative role of LET-23 such as a feedback inhibition mechanism.

Genetic mosaic analysis of the *let-23* gene has been performed to determine that the *let-23* gene function is required in the VPCs for vulval induction,[44,45] consistent with its function as the receptor that transduces the inductive signal. The results from the mosaic analysis also established that the lateral signal (interaction between an induced VPC and its neighboring VPCs) is sufficient to specify the 2° vulval fates. In mosaic animals, if the P6.p cell has wild-type *let-23* activity [*let-23(+)* in genotype]; but its two neighboring VPCs, P5.p and P7.p., have no *let-23* activity [*let-23(null)* in genotype], normal vulvae could usually be formed. That is, P5.p and P7.p could be induced to adopt the 2° vulval fate despite lacking the receptor for the inductive signal from the anchor cell. These experiments led to the proposal of a "sequential inductive model": the LIN-3 mediated anchor cell signal first induces the 1° fate in P6.p, and P6.p then produces a lateral signal that induces the 2° fate in the adjacent VPCs, P5.p and P7.p[44] (Fig. 2.1). The lateral signal appears to be mediated by the LIN-12 protein. Activated LIN-12 protein can promote the 2° vulval cell fate even if the *let-23 EGFR* or the *let-60 RAS* gene is mutated.[11,14,46]

The results from mosaic analysis of the *let-23* gene and the analysis of the dosage effect of *lin-3* transgenes strongly support the earlier proposal that the inductive signal and the lateral signal act together to specify the 2°-1°-2° cell pattern. The mechanism of how these two pathways integrate to specify cell fate is not clear at present.

SEM-5 AND *LET-341*: FACTORS BETWEEN EGFR AND RAS

As indicated by its name, *sem-5* mutations cause defects in sex-myoblast (SM) migration.[17] Several alleles of *sem-5* were isolated based on the phenotype of SM migration. However, three *sem-5* alleles reported by Clark et al[17] were isolated as suppressors of the Multivulva phenotype of a *lin-15* mutation. *sem-5* mutations also cause a Vulvaless phenotype and a partial larval lethal phenotype indicating the gene's essential function in vulval induction as well as in early larval development.

let-341 was initially defined by larval lethal mutations isolated in a screen for essential genes on chromosome V.[47] Like *sem-5*, its role in vulval induction was indicated by isolating mutations that suppress *lin-15* mutations.[18] The *let-341* mutations isolated in the suppressor screens cause incompletely penetrant larval lethality and many surviving animals display a Vulvaless phenotype.

Both *sem-5* and *let-341* appear to act between *let-23 EGFR* and *let-60 RAS* since mutations in *sem-5* and *let-341* suppress the Multivulva phenotype of *lin-15* but not the Multivulva phenotype of *let-60 RAS (gf)*.[18] The molecular study of *sem-5* indicated that it encodes a protein consisting almost entirely of one SH2 and two SH3 domains.[17] GRB-2, a mammalian homologue of SEM-5 was identified based on its ability to bind to a receptor tryosine kinase.[48] Functional homology between SEM-5 and GRB2 has also been demonstrated by showing that GRB2 as well as the *Drosophila* Drk gene can functionally replace *sem-5* in *C. elegans*.[49] The SH2 domain of GRB2 has been shown to interact with phosphotyrosine of receptor tyrosine kinases, while the SH3 domains of GRB2 have been shown to interact with SOS guanine nucleotide exchange factors.[50] Stern et al have also shown that recombinant SEM-5 protein can interact with phosphorylated, but not unphosphorylated, human EGF receptor. Such an interaction can be blocked by mutations in the SH2 domain.[49] Molecular isolation of the *let-341* gene has not yet been reported, although its position in the genetic pathway suggests it may encode a SOS-like guanine nucleotide exchange factor.

LIN-15 AND SYNTHETIC MULTIVULVA PHENOTYPE: THE INHIBITORY SIGNAL FROM THE HYPODERMIS

Recessive loss-of-function mutations in the *lin-15* locus cause a Multivulva phenotype, indicating that its wild-type gene products act negatively on the pathway. The *lin-15* locus actually consists of two genes, *lin-15A* and *lin-15B*, and only mutations affecting both genes result in a Multivulva phenotype.[5,51,52] The two *lin-15* genes belong to a group of genes called "synthetic Multivulva" genes.

The synthetic Multivulva phenotype was initially discovered from analyzing a mutant whose Multivulva phenotype results from an interaction between a mutation in *lin-8* and a mutation in *lin-9*.[53] Ferguson and Horvitz conducted screens to isolate additional mutations that would interact with either the *lin-8* or the *lin-9* mutation to generate the synthetic Multivulva phenotype (Fig. 2.2E).[5] Such screens identified a number of synthetic Multivulva genes that are divided into two classes, class A and class B. Class A genes include *lin-8, lin-38* and *lin-15A*, class B genes include *lin-9, lin-35, lin-36, lin-37* and *lin-15B*. Hermaphrodites carrying mutations in only one class have a wild-type vulva, but a combination of mutations in both A and B in the same strain confers a Multivulva phenotype. Ferguson and Horvitz[5] suggested that the A and B classes define two functionally redundant pathways that inhibit vulval cell fates. The synthetic Multivulva phenotype is epistatic to the Vulvaless phenotype caused by *lin-3* mutations but not that caused by *let-23* mutations, arguing that at least one of the two redundant inhibitory pathways acts upstream of the *let-23 EGFR* gene.

Genetic mosaic analyses of *lin-15* and *lin-37* led to the proposal that *lin-15* and other synthetic Multivulva genes are part of an inhibitory signal from the hypodermis to the EGF receptor on VPCs.[6,54] These experiments indicated that *lin-15* and *lin-37* do not act in the anchor cell or in the VPCs, but likely act in the syncytial hypodermis that surrounds the VPCs. The *lin-15A* and *lin-15B* genes have been cloned and they encode two novel proteins of unknown biochemical function. Because of the dramatic mutant phenotypes caused by these synthetic Multivulva mutations, further studies of the biochemical functions of these gene products will certainly reveal important aspects of EGF receptor-mediated signal transduction.

UNC-101, SLI-1 AND *ROK-1*: NEGATIVE REGULATORS OF EGFR SIGNALING

To identify new factors regulating *let-23 EGFR*-mediated signaling, Sternberg's laboratory screened for extragenic suppressors of the

Vulvaless phenotype of a severe loss-of-function but non-null mutation of *let-23*.[28,29] The *unc-101* and *sli-1* genes were identified by some of these suppressor mutations. Both genes act negatively on the signaling pathway because loss-of-function mutations in these genes overcome reduction of function in the *let-23 EGFR* gene. Neither *unc-101* nor *sli-1* mutations cause a defect in vulval cell differentiation on their own, but the *unc-101; sli-1* double mutants display a Multivulva phenotype. Since *unc-101* and *sli-1* do not display strong synergism with *lin-15A* or *lin-15B*, they appear to represent distinct negative activities.[55] Genetic analysis also suggests that both genes act negatively at or near the *let-23*-mediated step of the vulval induction pathway.[28,29] Since *unc-101* and *sli-1* mutations do not suppress null mutations of *let-23*, they do not bypass the *let-23* gene function.

unc-101 appears to play roles in multiple aspects of development and behavior since mutations in the gene cause pleiotropic defects including uncoordinated movement, irregularity in defecation, and abnormal development of the male tail.[28,56] In fact, *unc-101* was first identified by mutations causing uncoordinated movement due to defects in neuronal development.

unc-101 encodes a homologue of the mammalian AP47 protein, a component of the trans-Golgi adaptin complex.[28] UNC-101 protein is about 70% identical over 422 amino acids to mouse AP47 and can be functionally replaced by the mouse AP47 transgene in *C. elegans*. Adaptins link transmembrane proteins such as the EGF receptor to clathrin and can drive clathrin coat assembly onto membranes in vitro.[57] The genetic function and the molecular identity of *unc-101* suggest a role for clathrin-coated vesicles in regulating the activity of the LET-23 EGFR; such a role might explain recent findings that a human meningioma tumor suppressor locus encodes a β adaptin.[58]

sli-1 encodes a protein related to the product of the mammalian *cbl* proto-oncogene.[60] c-Cbl is a cytoplasmic protein containing a C3HC4 Zinc-binding motif known as the RING finger motif and several potential SH3-binding sites.[59] SLI-1 and c-Cbl share about 55% amino acid identity over a 390 residue region containing these motifs. The cellular function of c-Cbl is not clear, though deletions of the C-terminus including the RING finger are oncogenic.[59] *sli-1* mutations can suppress the Vulvaless phenotype caused by severe reduction-of-function mutations of *sem-5* or *let-23*.[29] One of these *let-23* alleles alters the splicing of the *let-23* transcript, and results in a truncation of the C-terminal sequence containing the likely binding sites for the SEM-5 protein.[43] Yoon et al[60] thus speculated that the normal function of SLI-1 may be to limit interactions between the LET-23 EGFR and SEM-5 or to suppress an alternative (SEM-5 independent) pathway of signal transduction from the LET-23 EGFR to LET-60 RAS.

Because double mutants defective in both *unc-101* and *sli-1* display a synergistic Multivulva phenotype that is not associated with single mutations in either gene, Sternberg and colleagues have also screened for mutations that cause a Multivulva phenotype in a *sli-1* mutant background to identify additional regulatory genes. Such a screen recovered additional mutations in *unc-101, lin-2* and *lin-10* as well as mutations in a new gene called *rok-1* (*r*egulator *o*f *k*inase-mediated signaling).[55] The *rok-1* mutation results in a temperature sensitive Multivulva phenotype and can also cause a Multivulva phenotype when combined with a mutation of either *sli-1* or *unc-101*. Furthermore, a triple mutant defective in all three genes has a more severe Multivulva phenotype than do any of the three double mutant combinations. Further molecular study of *rok-1* may shed light on the biochemical function of this new negative regulator.

LIN-2, 7 AND *10:* POSITIVE REGULATORS OF EGFR SIGNALING

The *lin-2, lin-7* and *lin-10* genes are defined by recessive loss-of-function mutations that cause a Vulvaless phenotype.[12,13] Since they share similar mutant phenotypes and other genetic properties, these three genes may function at a common step in the signaling process. For example, the defects caused by mutations of these genes (either singly or in combination) seem to be limited to vulval development, suggesting that they are positive regulators specific to vulval induction. Null mutations in any of the three genes or combinations of mutations in all three genes do not completely eliminate vulval induction, suggesting that they may act as modifiers rather than as the essential components of the "main" signaling pathway. Genetic epistasis experiments using a *let-23 EGFR* gain-of-function allele showed that these three genes act upstream of *let-23 EGFR.*[39] Genetic mosaic analysis of the *lin-2* and *lin-7* genes indicated that both genes act in the VPCs.[44,61]

lin-2 encodes a member of the membrane-associated guanylate kinase (MAGUK) family of cell junction proteins.[61] MAGUK proteins contain three common domains: a GLGF (Gly-Leu-Gly-Phe) motif that mediates protein binding, a single SH3 domain and a single guanylate kinase region.[62] MAGUK proteins have been implicated in the maintenance or establishment of epithelial cell polarity. The N-terminus of the LIN-2 protein also contains a region that is similar to Ca^{2+}/calmodulin-dependent protein kinase II (Ca kinase II). Because mutant *lin-2* transgenes lacking either protein kinase or guanylate kinase activity are still functional, LIN-2 is likely to have a structural rather than an enzymatic role in vulval induction. Specifically, Hoskin et al[61] have proposed that the LIN-2 protein is required for proper localization of the LET-23 EGFR or other signaling molecules to either the basal membrane domain or the cell junctions of VPCs. *lin-7* encodes a protein containing a single GLGF repeat which may suggest that its function is related to that of the *lin-2* gene.[62] *lin-10* encodes a novel protein.[63]

V. GENES ACTING DOWNSTREAM OF RAS

LIN-45 RAF, *MEK-2* AND *SUR-1/MPK-1*: A CONSERVED KINASE CASCADE

The roles of *lin-45*, *mek-2*(also known as *let-537*) and *sur-1/mpk-1* in the vulval signaling pathway were determined by analyzing chromosomal mutations that cause a Vulvaless phenotype and/or suppress the Multivulva phenotype of existing mutations.[19-23,64] Mutations in all three genes suppress the Multivulva phenotype of the *let-60 RAS(G13E gf)* allele, establishing their functional position downstream of RAS in the pathway. Because of the high degree of similarity between the proteins encoded by these three genes and their mammalian counterparts, the Raf, Mek and MAPK proteins,[19-22] the order of action of these three genes (Fig. 2.3) is proposed based on the functional relationship established for the mammalian kinases. The functional homology between the worm genes and mammalian genes has been demonstrated by two lines of experiments. First, a mouse MAP kinase transgene can rescue the mutant phenotype of a mutation in the *C. elegans sur-1* gene.[21] Second, the *C. elegans* MEK-2 protein has been shown to phosphorylate and activate a mammalian MAP kinase, and MEK-2 itself can be phosphorylated and activated by mammalian Raf-1 in vitro.[22]

The *lin-45* gene was initially defined by two partial loss-of-function mutations isolated as suppressors of *lin-15* mutations.[14,18] Additional mutations were isolated as suppressors of an activated *let-60 RAS* mutation, G13E[27] (unpublished results from our laboratory; Kornfeld and Horvitz, personal communication). Because null mutations and mutations with a completely penetrant Vulvaless phenotype have not been reported, it has not been determined if *lin-45* is absolutely required for vulval induction as are *let-60 RAS* and *let-23 EGFR*. It is possible that the pathway branches after *let-60 RAS*, as suggested by studies on the *ksr-1* gene (see below).

Several recessive mutations in the *mek-2* and *sur-1/mpk-1* genes were isolated as suppressors of activated *let-60 ras*[20-23] (unpublished results from our laboratory). Additional mutations in *mek-2* and *sur-1/mpk-1* were isolated based on their effects on germline development or fertility.[64,65] Many *mek-2* mutations including a putative null mutation cause a recessive Vulvaless phenotype as well as recessive sterility,[23,64] presumably due to defects in germline meiotic cell cycle progression described by Church et al.[64] Existing mutations in the *sur-1/mpk-1* gene cause only mild defects. Of the two initially reported mutations, one causes only a partial vulval lineage defect and low percentage of larval lethality,[21] and the other causes no obvious defects.[20] A third allele causes a severe sterile phenotype similar to that of many *mek-2* alleles.[64] The null phenotype of *sur-1/mpk-1* is not clear. Attempts to isolate null alleles by noncomplementation screens using existing alleles have so far not been fruitful (Y. Wu and M. Han, unpublished; Lackner and Kim, personal communications). It is possible that null mutations are lethal in trans to an existing allele so that they would not be recovered in the screens. Mosaic analysis of *sur-1/mpk-1* showed that the gene acts in the VPCs to transduce the inductive signal.[20]

KSR-1: A Positive Modifier of RAS-Mediated Signaling

The *ksr-1* (*k*inase *s*uppressor of *R*AS) gene appears to be required for efficient or maximal signaling output downstream of *let-60 RAS*. It is defined by ten mutations isolated as suppressors of *let-60 RAS(G13E gf)* mutations.[24,27] Two additional mutations were isolated in a noncomplementation screen intended to identify null mutations.[27] *ksr-1* mutations do not cause obvious defects in vulval cell lineages in an otherwise wild-type background, but do cause several other pleiotropic defects including partial larval lethality and abnormal sex-myoblast migration. *ksr-1* mutations cause strong synergistic Vulvaless and lethal phenotypes in combination with weak alleles of *lin-45 Raf* or *sur-1/mpk-1*, suggesting that loss of *ksr-1* reduces signaling activity mediated by wild type *let-60 RAS*.[24,27] Extensive genetic characterization and molecular sequencing of these *ksr-1* mutations indicate that they are likely loss-of-function mutations.

ksr-1 encodes a novel protein kinase that is most closely related to the Raf Ser/Thr kinase family.[24,27] The homology is about 35% in the kinase domain between KSR-1 and mammalian Raf-1. Like Raf, the N-terminal half of the KSR-1 protein also contains a cysteine-rich zinc finger domain. However, KSR-1 does not appear to be a kinase of the Raf family since the resemblance between KSR-1 and Raf is relatively low compared to that among different Raf proteins. The N-terminal half of KSR-1 does not contain the RAS binding site I and the serine/threonine-rich CR2 motif present in Raf proteins. KSR-1 does seem to contain the sequence similar to RAS-binding site II within the Cys-rich domain of Raf (see chapter 5), raising the possibility that KSR-1 might still bind RAS through this domain. However, interaction between KSR-1 and LET-60 RAS was not detected in experiments using the yeast two-hybrid system (M.S. and M.H., unpublished observation). It is also not clear whether KSR-1 protein is a serine/threonine kinase or a tyrosine kinase since it contains structural features that resemble both types of kinases. The *Drosophila Ksr* gene, which encodes a protein similar to the *C. elegans ksr-1* gene, has also been identified by isolating mutations that suppress an activated Ras1 mutant.[66] Therrien et al also reported the cloning of KSR homologues in *Drosophila virillis*, mouse and human, suggesting that both the structure and the function of KSR proteins may be conserved among different organisms.

At present, the mechanism by which KSR proteins influence the RAS pathway is not known. Genetic epistasis tests in *C. elegans* and *Drosophila* suggest that they act either downstream of or in parallel to RAS, but upstream of Raf.[24,27,66] One of many possibilities is that KSR proteins mediate a branch of the pathway downstream of RAS but parallel to Raf.

LIN-1 AND *LIN-31*: TRANSCRIPTION FACTORS ACTING LATE IN THE PATHWAY

The *lin-1* and *lin-31* genes encode putative transcription factors acting downstream of *sur-1/mpk-1* in the pathway. Loss-of-function mutations in *lin-1* cause a Multivulva phenotype, indicating that *lin-1* acts negatively in the pathway.[10,13,67] *lin-1* mutations suppress the Vulvaless and lethal phenotypes associated with reduced *let-60 RAS* or *sur-1/mpk1* activity, suggesting that its gene product acts after MAP kinase. However, in *lin-1* null mutants, VPCs still respond to signaling from the gonadal anchor cell, indicating that other factors must also act to transduce the anchor cell signal in a *lin-1*-independent manner.[67] It is possible that *lin-1* may act in only one of two or several branches of the RAS-mediated signaling pathway.[25,26,67] *lin-1* encodes an ETS-domain protein that contains several potential MAPK phosphorylation sites, suggesting that it may be a direct target of MAP kinase.[67] However, a requirement for these sites and phosphorylation of LIN-1 protein by SUR-1/MPK-1 have not been demonstrated.

The *lin-31* mutant phenotype is significantly different from that of other genes acting in the signaling pathway.[68] Loss-of-function mutations in *lin-31* appear to cause the fate of the six VPCs to be deregulated: the three VPCs (P3.p, P4.p and P8.p) that are normally not induced often adopt vulval fates, while the three VPCs that are normally induced often adopt nonvulval hypodermal cell fates. It has also been shown that VPCs can still adopt vulval cell fates in *let-60 RAS (dn); lin-31* double mutants, suggesting that vulval fates in a *lin-31* mutant do not require RAS pathway activity.[68] Double mutant analysis also suggests that *lin-31* may act downstream of MAP kinase.[20] *lin-31* encodes a protein that resembles the HNF-3/fork head family of DNA-binding transcription factors containing acidic and proline-rich regions.[68] Because the LIN-31 protein also contains a potential MAPK phosphorylation site, it is also a candidate target of SUR-1/MPK-1. Miller et al[68] have proposed that LIN-31 acts late in the signaling pathway to regulate how VPCs choose their fates; LIN-31 may form different complexes or exist in different forms that can repress vulval cell fates or promote hypodermal cell fates in some VPCs, but activate vulval fates or repress hypodermal fates in other VPCs. Phosphorylation of LIN-31 by MAPK may a play role in generating different forms of LIN-31. According to this model, cells that do not contain any LIN-31 are deregulated and adopt fates at random.

SUR-2 AND *LIN-25*: NOVEL GENES ACTING LATE IN THE SIGNALING PATHWAY

Mutations in *sur-2* and *lin-25* cause similar phenotypes, and both genes appear to act as positive factors later in the signaling pathway.[25,26] Loss-of-function mutations in *sur-2* or *lin-25* cause an incompletely penetrant Vulvaless phenotype and other pleiotropic defects including partial larval lethality, sterility and defective male tail development. Many *sur-2* and *lin-25* mutations were isolated as suppressors of the Multivulva phenotype of *let-60(G13E gf)*, suggesting both genes act after RAS in the pathway. The genetic interactions among *lin-1, lin-31, lin-25* and *sur-2* are complicated and do not support a simple linear ordering of their activities, but do suggest that both *sur-2* and *lin-25* act late in the pathway, likely after MAPK.[25,26] For instance, *lin-25* and *sur-2* mutations each can suppress the Multivulva phenotype caused by a weak *lin-1* allele but not that caused by a *lin-1* null allele. The deregulated fates of P(3,4,8).p in *lin-31* mutants are suppressed by a *sur-2* mutation, suggesting that SUR-2 may be negatively regulated by LIN-31 in these cells. However, the *lin-31* mutant phenotype of P(5-7).p is not suppressed by the *sur-2* mutation, suggesting *that sur*-2 does not act downstream of *lin-31* in these three cells. It is possible that *lin-1, lin-31, sur-2* and *lin-25* act in parallel branches downstream of MAPK with some cross talking among them. A *sur-2::lacZ* reporter gene is expressed in VPCs at the time of vulval determination, consistent with its function in vulval induction.[25]

VI. MULTIPLE FUNCTIONS OF THE *LET-60 RAS*-MEDIATED SIGNAL TRANSDUCTION PATHWAY DURING *C. ELEGANS* DEVELOPMENT

As discussed above, mutations in most of the genes involved in vulval signaling cause multiple developmental defects. For example, severe loss-of-function mutations in *lin-3, let-23, sem-5, let-60, lin-45, mek-2, sur-1/mpk-1, ksr-1, sur-2* and *lin-25* may result in larval lethality, sterility, and/or male tail defects. The severity of these pleiotropic phenotypes depends on the specific allele. Similar phenotypes associated with some of the genes suggest that some key factors in the pathway may act together as a unit in multiple developmental events. For example, analyses of the male tail defects associated with many genes, including *lin-3, lin-15, let-23, sem-5, let-60* and *lin-45*, indicated that these genes act together in a similar pathway to transduce signals that specify certain cell fates in the male tail.[69] By analyzing the sterile phenotype associated with certain mutations in the *let-60 RAS, mek-2* and *sur-1/mpk-1* genes, Chuch et al[64] also revealed the common roles of these three genes in meiotic cell cycle progression. Mutations in these three genes cause a failure of meiotic germ cells to exit from pachytene to produce oocytes and sperm. Genetic double mutant analysis showed that *lin-1* does not act downstream of *sur-1/mpk-1* to specify such a function in germline cells. Because certain developmental events (particularly those occurring in later postembryonic stages such as vulval differentiation and male tail development) are more sensitive to reduction of gene activity in the RAS pathway than are other events, many other functions of this pathway in development are yet to be learned. From all the studies done in different organisms (including those using mammalian tissue culture cells), it is now clear that the RAS-mediated signal transduction pathway plays important roles in cell proliferation, differentiation, migration, and behavior responses.

What controls the ability of these genes to trigger specific developmental events in different cell types? Specificity must be provided by other cellular factors which could act at the beginning, the end, and possibly several other points of this common pathway. For example, factors are likely to act in the anchor cell to control specific expression of the LIN-3 signal so that it can only activate LET-23 EGFR during a specific period of time. Other specific factors must also act downstream of MAP kinase to connect the signal transduction pathway with specific functions such as vulval differentiation and germ cell exit from pachytene. It has recently become clear, particularly from studies in mammalian tissue culture cells, that the RAS-mediated signaling pathway is hardly a single, linear cascade. Instead, different signaling pathways or complexes are connected by a signaling network (see other chapters in this book). Cross-talk between different pathways may vary from tissue to tissue and from one developmental event to another, and thus may provide additional regulation for specific functions. One possible reason for eukaryotic cells to use a long signal transduction pathway such as the RTK/RAS/MAPK pathway is that each of the components in the pathway could interact with other factors and thus allow diverse mechanisms of regulation. Comparisons of RAS pathway functions in different developmental events should be important and may provide significant information about the regulation and specificity of this commonly used signaling pathway.

Acknowledgments

M.S. is a Boehringer Mannheim Fellow of the Life Sciences Research Foundation. M.H. is a Lucille P. Markey Scholar and a Searle Scholar. Research in M.H.'s laboratory is also supported by grants from the United States Public Health Service (R01 GM47869), the Council for Tobacco Research and the March of Dimes Foundation.

References

1. Sulston J, Horvitz HR. Postembryonic cell lineages of the nematode *Caenorhabditis elegans*. Devel Biol 1977; 56:110-56.
2. Sundaram M, Han M. Control and integration of cell signaling pathways during *C. elegans* vulval development. BioEssay 1996; 18:473-480.

3. Sternberg PW. Intercellular signaling and signal transduction in *C. elegans*. Annu Rev Genet 1993; 27:497-521.
4. Katz WS, Hill RJ, Clandinin TR et al. Different levels of the *C. elegans* growth factor LIN-3 promote distinct vulval precursor fates. Cell 1995; 82:297-307.
5. Ferguson E, Horvitz HR. The multivulva phenotype of certain *C. elegans* mutants results from defects in two functionally-redundant pathways. Genetics 1989; 123:109-21.
6. Herman RK, Hedgecock EM. Limitation of the size of the vulval primordium of *Caenorhabditis elegans* by *lin-15* expression in surrounding hypodermis. Nature 1990; 348:169-71.
7. Greenwald I, Rubin GM. Making a Difference: the role of cell-cell interactions in establishing separate identities for equivalent cells. Cell 1992; 68:271-81.
8. Horvitz HR, Sternberg PW. Multiple intercellular signaling systems control the development of the *C. elegans* vulva. Nature 1991; 351:535-41.
9. Sternberg PW. Intercellular signaling and signal transduction in *C. elegans*. Ann Rev Genet 1993; 27:497-521.
10. Horvitz HR, Sulston JE. Isolation and genetic characterization of cell-lineage mutants of the nematode *Caenorhabditis elegans*. Genetics 1980; 96:435-54.
11. Greenwald IS, Sternberg PW, Horvitz HR. The *lin-12* locus specifies cell fates in *Caenorhabditis elegans*. Cell 1983; 34:435-44.
12. Ferguson E, Horvitz HR. Identification and characterization of 22 genes that affect the vulval cell lineages of Caenorhabditis elegans. Genetics 1985; 110:17-72.
13. Ferguson EL, Sternberg PW, Horvitz HR. A genetic pathway for the specification of the vulval cell lineages of *Caenorhabditis elegans*. Nature 1987; 326:259-67.
14. Han M, Aroian R, Sternberg PW. The *let-60* locus controls the switch between vulval and non-vulval cell types in *C. elegans*. Genetics 1990; 126:899-913.
15. Beitel G, Clark S, Horvitz HR. The *Caenorhabditis elegans ras* gene *let-60* acts as a switch in the pathway of vulval induction. Nature 1990; 348:503-9.
16. Aroian RV, Sternberg PW. Multiple functions of *let-23*, a *Caenorhabditis. elegans* receptor tyrosine kinase gene required for vulval induction. Genetics 1991; 128:251-67.
17. Clark SG, Stern MJ, Horvitz HR. *C. elegans* cell-signaling gene *sem-5* encodes a protein with SH2 and SH3 domain. Nature 1992; 356:340-4.
18. Clark SG, Stern MJ, Horvitz HR. Genes involved in two *Caenorhabditis elegans* cell-signaling pathways. Cold Spring Harbor Symposia on Quantitative Biology 1992; LVII: 363-73.
19. Han M, Golden A, Han Y et al. *C. elegans lin-45 raf* gene participates in *let-60 ras*-stimulated vulval differentiation. Nature 1993; 363:133-40.
20. Lackner MR, Kornfeld K, Miller LM et al. A MAP kinase homolog,

mpk-1, is involved in ras-mediated induction of vulval cell fates in *C. elegans*. Genes Dev 1994; 8:160-73.

21. Wu Y, Han M. Suppression of activated Let-60 RAS protein defines a role of Caenorhabditis elegans sur-1 MAP kinase in vulval differentiation. Genes Dev 1994; 8:147-59.
22. Wu Y, Han M, Guan KL. MEK-2, a *Caenorhabditis elegans* MAP kinase kinase, function in RAS-mediated vulval induction and other developmental events. Genes & Development 1995; 9:742-55.
23. Kornfeld K, Guan KL, Horvitz HR. The *C. elegans* gene mek-2 is required for vulval induction and encodes a protein similar to the protein kinase MEK. Gen & Devel 1995; 9:757-67.
24. Kornfeld K, Hom DB, Horvitz HR. The ksr-1 gene encodes a novel protein kinase involved in RAS-mediated signaling in Caenorhabditis elegans. Cell 1995b; 902-13.
25. Singh N, Han M. *sur-2*, a novel gene, functions late in the *let-60 ras*-mediated signaling pathway during *Caenorhabditis elegans* vulval induction. Gen & Devel 1995; 9:2251-65.
26. Tuck S, Greenwald I. lin-25, a gene required for vulval induction in Caenorhabditis elegans. Genes & Dev 1995; 9:341-57.
27. Sundaram M, Han M. The *C. elegans* ksr-1 gene encodes a novel Raf-related kinase involved in RAS-mediated signal transduction. Cell 1995; 83:889-901.
28. Lee JG, Jongeward G, Sternberg PW. The *C. elegans* unc-101 gene, required for development and behavior, encodes a clathrin-associated protein. Genes Devel 1994; 8:60-73.
29. Jongeward GD, Clandinin TR, Sternberg PW. *sli-1*, a negative regulator of *let-23*-mediated signaling in *C. elegans*. Genetics 1995; 139:1553-66.
30. Trent C, Tsung N, Horvitz HR. Egg-laying defective mutants of the nematode *Caenorhabditis elegans*. Genetics 1983; 104:619-47.
31. Wassarman DA, Thierrien M, Rubin GM. The RAS signaling pathway in *Drosophila*. Curr Opin Gen Dev 1995; 5:44-51.
32. Rogalski TM, Moerman DG, Baillie DL. Essential genes and deficiencies in the unc-22 IV region of Caenorhabditis elegans. Genetics 1982; 102:725-36.
33. Clark DV, Rogalski TM, Donati LM et al. The *unc*-22(IV) region of Caenorhabditis elegans: genetic analysis of lethal mutations. Genetics 1988; 119:345-53.
34. Howe LR, Marshall CJ. Identification of amino acids in P21 ras involved in exchange factor interaction. Oncogene 1993; 8:2583-90.
35. Han M, Sternberg PW. *let-60*, a gene that specifies cell fates during *C. elegans* vulval induction, encodes a ras protein. Cell 1990; 63:921-31.
36. Han M, Sternberg PW. Analysis of dominant-negative mutations of the *Caenorhabditis elegans let-60 ras* gene. Genes Dev 1991; 5:2188-98.
37. Sigal IS, Gibbs JB, D'Alonzo JS et al. Mutant *ras*-encoded proteins with altered nucleotide binding exert dominant biological effects. Proc Natl Acad Sci USA 1986; 83:952-6.

38. Hill RJ, Sternberg PW. The gene *lin-3* encodes an inductive signal for vulval development in *C. elegans*. Nature 1992; 358:470-6.
39. Katz WS, Lesa GM, Yannoukakos D et al. A point mutation in the extracellular domain activates LET-23, the *C. elegans* EGF receptor homolog. Mole Cell Biol 1996; (in press).
40. Herman RK. Crossover suppressors and balanced recessive lethals in *Caenorhabditis elegans*. Genetics 1978; 88:49-65.
41. Sigurdson DC, Spanier GJ, Herman RK. *Caenorhabditis elegans* deficiency mapping. Genetics 1984; 108:331-45.
42. Aroian RV, Koga M, Mendel JE et al. The *let-23* gene necessary for *Caenorhabditis elegans* vulval induction encodes a tyrosine kinase of the EGF receptor subfamily. Nature 1990; 348:693-9.
43. Aroian RV, Lesa GM, Sternberg PW. Mutations in the Caenorhabditis elegans let-23 EGFR-like gene define elements important for cell-type specificity and function. EMBO J 1994; 13:360-6.
44. Simske JS, Kim SK. Sequential signaling during *C. elegans* vulval induction. Nature 1995; 375:142-6.
45. Koga M, Ohshima Y. Mosaic analysis of the let-23 gene function in vulval induction of Caenorhabditis elegans. Devel 1995; 121:2655-66.
46. Sternberg PW, Horvitz HR. The combined action of two intercellular signaling pathways specifies three cell fates during vulval induction in *C. elegans*. Cell 1989; 58:679-93.
47. Johnsen RC, Baillie DL. Genetic analysis of a major segment [LGV(left)] of the genome of *Caenorhabditis elegans*. Genetics 1991; 129:735.
48. Lowenstein EJ, Daly RJ, Batzer AG et al. The SH2 and SH3 domain-containing protein GRB2 links receptor tyrosine kinases to ras signaling. Cell 1992; 70:431-42.
49. Stern MJ, Marengere LEM, Daly RJ et al. The human GRB2 and Drosophila Drk genes can functionally replace the Caenorhabditis elegans cell signaling gene sem-5. Mol Biol Cell 1993; 4:1175-88.
50. McCormick F. How receptors turn RAS on. 1993; 363:15-6.
51. Huang LS, Tzou P, Sternberg PW. The *lin-15* locus encodes two negative regulators of *Caenorhabditis elegans* vulval development. Molecular Biology of the Cell 1994; 5:395-411.
52. Clark SG, Lu X, Horvitz HR. The *Caenorhabditis elegans* locus *lin-15*, a negative regulator of a tyrosine kinase signaling pathway, encodes two different proteins. Genetics 1994; 137:987-97.
53. Sulston JE, Horvitz HR. Abnormal cell lineages in mutants of the nematode *Caenorhabditis elegans*. Dev Biol 1981; 82:41-55.
54. Hedgecock EM, Herman RK. The ncl-1 gene and genetic mosaics of Caenorhabditis elegans. Genet 1995; 141:989-1006.
55. Sternberg PW, Yoon CH, Lee J et al. Molecular genetics of proto-oncogenes and candidate tumor suppressors in Caenorhabditis elegans. Cold Spring Harbor Symp Quant Biol 1994; LIX: 155-63.
56. Thomas JH. Genetic analysis of defecation in *Caenorhabditis elegans*. Genetics Society of America 1990; 124:855-72.

57. Schmid SL. The mechanism of receptor-mediated endocytosis. BioEssays 1992; 14:589-96.
58. Peyrard M, Fransson I, Xie YG et al. Characterization of a new member of the human B-adaptin gene family from chromosome 22q12, a candidate meningioma gene. Hum Mol Genet 1994; 3:1393-9.
59. Blake TJ, Shapiro M, Morse HC et al. The sequences of the human and mouse c-*cbl* proto-oncogenes show v-cbl was generated by a large truncation encompassing a proline-rich doman and a leucine zipper-like motif. Oncogene 1991; 6:653-.
60. Yoon CH, Lee J, Jongeward GD et al. Similarity of sli-1, a regulator of vulval development in *C. elegans*, to the mammalian proto-oncogene c-cbl. Sci 1995; 269:
61. Hoskins R, Hajnal AF, Harp SA et al. The *C. elegans* vulval induction gene lin-2 encodes a member of the MAGUK family of cell junction proteins. Devel 1996; (in press).
62. Kim SK. Tight junctions, membrane-associated guanylate kinases and cell signaling. Curr Opin Cell Biol 1995; 7:641-9.
63. Kim SK, Horvitz HR. The *Caenorhabditis elegans* gene *lin-10* is broadly expressed while required specifically for the determination of vulval cell fates. Genes & Devel 1990; 4:357-71.
64. Chuch DL, Guan KL, Lambie EJ. mek-2, mpk-1/sur-1 and let-60 ras are required for meiotic cell cycle progression in *C. elegans*. Develop 1995; 121:2525-35.
65. McKim KS. (The University of British Columbia, Vancouver, British Columbia, Canada., 1990).
66. Therrien M, Chang HC, Solomon NM et al. KSR, a novel protein kinase required for RAS signal transduction. Cell 1995; 83:879-87.
67. Beitel GJ, Tuck S, Greenwald IS et al. The *C. elegans* gene *lin-1* encodes an ETS-domain protein and defines a branch of the vulval induction pathway. Genes Dev 1996; 9:3149-62.
68. Miller LM, Gallegos ME, Morisseau BA et al. Lin-31, a Caenorhabditis elegans HNF-3/fork head transcription factor homolog, specifies three alternative cell fates in vulval development. Genes & Dev 1993; 7:933-47.
69. Chamberlin HM, Sternberg PW. The *lin-3/let-23* pathway mediates inductive signaling during male speicule development in *Caenorhabditis elegans*. Development 1994; 120:2713-21.
70. Sulston JE. Post-embryonic development in the ventral cord of *Caenorhabditis elegans*. Philos Trans R Soc Lond B Biol Sci 1976; 275:287-98.
71. Sulston JE, White JG. Regulation and cell autonomy during postembryonic development of *Caenorhabditis elegans*. Devel Biol 1980; 78:577-97.
72. Sternberg PW, Horvitz HR. Pattern formation during vulval development in *Caenorhabditis elegans*. Cell 1986; 44:761-72.
73. Kimble J. Lineage alterations after ablation of cells in the somatic gonad of *Caenorhabditis elegans*. Dev Biol 1981; 87:286-300.
74. Sternberg PW. Lateral inhibition during vulval induction in *Caenorhabditis elegans*. Nature 1988; 335:551-4.

CHAPTER 3

Mammals I: Regulation of RAS Activation

Antony W. Burgess

I. INTRODUCTION

The discovery of the RAS oncogene was one of the pivotal events in modern cancer research.[1-3] However, the importance of the discovery was in many ways due to our knowledge of hormone action and the role of G-proteins in signaling.[4] v-Ha-RAS was known to bind GTP and become phosphorylated when the GTP was hydrolyzed to GDP.[5] The similarity between the possible actions of RAS-GTP and other members of the G-protein system was noticed soon after the identification of c-Ha-RAS and its ability to hydrolyze GTP.[4-6]

The notion that RAS acted as an amplification switch for growth factor and cytokine signaling gradually emerged as our knowledge of cell growth control and signaling improved. Soon the centrality of the G-protein-GDP→G-protein-GTP switch in the mediation of hormone action dominated much of signaling research and many models of hormone action.[7] However, the molecules controlling the RAS switch and the targets for activated RAS* have been most elusive. Further progress in our understanding of RAS activation came not from cancer research per se, but from some basic studies about the genes responsible for eye development in fruit flies (*Drosophila melanogaster*)[8] and vulval development in *C. elegans.*[9] The regulatory pathway for signaling from a particular tyrosine kinase receptor system through RAS and Raf was mapped using the *Drosophila* mutants. A regulator of GTP loading of RAS was identified by mutational analysis:[10] this regulator

Regulation of the RAS Signaling Network, edited by Hiroshi Maruta and Antony W. Burgess.

was called Son-of-Sevenless (SOS). SOS was required to catalyze the removal of GDP from RAS and consequentially facilitate the binding of GTP to RAS. Once in the GTP-form, RAS* interacts with many powerful proteins involved in cell morphology,[11] membrane ruffling,[12] cellular communication[13] and mitosis.[14] Indeed, the array of proteins interacting with RAS*-GTP is so large it is still difficult to identify the critical events associated with a specific signaling pathway.[15]

The identification of SOS, and the realization that when many cell surface receptors were activated by their appropriate ligand the level of intracellular tyrosine kinase increased,[16] has allowed a much better understanding of the signaling pathways from receptor to the formation of GTP-RAS. In particular, the importance of the phosphorylation of individual tyrosines on receptors[17,18] and shuttle molecules,[19] as well as the recognition of proteins with domains which bind these phosphotyrosine motifs[20,21] has led to major developments in our understanding of the intracellular events which link extracellular ligand binding to the activation of the RAS complex. The regulation of phosphotyrosine binding sites (by the kinases and phosphatases), binding by SH2 (and other motifs), and the linkage of proteins in arrays by adaptors with multiple binding motifs (e.g., Grb2) provide a simple model for the coordination of signals:

$$R \rightarrow [L{\cdot}R^*(YP)_n]_m \rightarrow [L{\cdot}R^*(YP)_n]_m{\cdot}Grb2{\cdot}SOS^* \rightarrow [L{\cdot}R^*(YP)_n]_m{\cdot}Grb2{\cdot}SOS^*RAS{\cdot}GDP$$
$$\rightarrow [L{\cdot}R^*(YP)]_m{\cdot}Grb2{\cdot}SOS^*{\cdot}RAS \rightarrow [L{\cdot}R^*(YP)_n]_m{\cdot}Grb2{\cdot}SOS^*{\cdot}RAS^*{\cdot}GTP$$

Where R = receptor, L = ligand, * = active state, YP = phosphotyrosine, n = number of phosphorylation sites and m = oligomerization state of the receptor.

The connection between the receptor and RAS may not always be this direct; indeed, this process may not be the major signaling pathway for many receptors, but elements of (and variations on) this pathway appear to underpin the early mitogenic events triggered by many growth factors and/or cytokines. Other kinases, adaptors (e.g. SHC),[19] guanine nucleotide exchange factors[22] (e.g. RAS-GRF[23] [also called CDC25^{Mm}][24]) may be involved in specific cell types, but the general concept of:

$$\begin{array}{ccccccccc} R & \rightarrow & R^* & \rightarrow & A^* & \rightarrow & GEF^* & \rightarrow & RAS^* \\ & & \downarrow & & & & & & \downarrow \searrow \\ & & ? & & & & & & \end{array}$$

where * = activated state, R = Receptor, A = adaptor, and ?= other signaling systems, appears to be relevant to most signaling systems. The specific proteins which take part in a particular RAS activation pathway will depend on the differentiation state of the cell and the growth factor/cytokine involved in the signaling process.

For many of the growth factor/cytokine signal transduction pathways the activation of RAS* is an essential step in the biological activation process (e.g., mitosis or gene activation). Blocking signaling from RAS*-GTP with antibodies[25] or prevention of the formation of RAS*-GTP with a dominant negative mutant of RAS (e.g., $N_{17}RAS$)[26] invariably blocks signal transduction. However, it must be remembered that other processes upstream of RAS are required for signaling by growth factors. The nature and role of the other signaling pathways (e.g. JAK/STAT[27-29] or PI3 Kinase[30]) and their relationship with the RAS-activated pathways are under intense investigation at present.

Another important aspect of RAS-related signaling concerns the range of target genes activated by RAS*. The RAS*→Raf1--->MAPK,[31] RAS*↔GAP,[32] RAS*↔Rac1,[12,33] RAS*↔PI3K,[30] RAS*→PKC-zeta and/or RAS*↔NF1[34] interactions are capable of stimulating hundreds of different cellular processes; however, often RAS* also leads to the induction of growth factor and cytokine genes (e.g., TGFα[35] or IL-2[36]). Indeed, the initial discovery of autocrine growth factors and their role in tumorigenesis came from studies on v-RAS transformed fibroblasts.[13,35] It was clear that transformation by v-RAS regulated-down the display of the epidermal growth factor receptor.[13] Subsequently, it was found that this was caused by autocrine secretion of TGF–α,[35] which bound, activated and induced the internalization/degradation of the EGF receptor. Whilst the induction of growth factor/cytokine secretion appears to be important for the action of oncogenic RAS,[37] the role of growth factor production induced by RAS activation during normal cellular processes is not yet as clear. When a single growth factor stimulates a cell, the multiple processes required for mitosis may require activation of other cell surface receptors. This could be achieved by linking the expression of tissue-specific growth factors to the RAS*-GTP signaling pathway. These RAS-dependent autocrine growth factors would not only stimulate the extra signaling pathways, but would also amplify the formation of RAS*-GTP. Conceptually, this would mean RAS* controls both upstream and downstream signaling processes. In activating the upstream (receptor-driven) processes, oncogenic RAS* could be expected to have biological effects usually associated with specific ligand-induced processes (e.g., proliferation and/or differentiation).[38] In this chapter, some of the receptor/kinase systems linked to RAS will be considered. The variations on the general theme and the specific molecular interactions responsible for linking different receptor systems to RAS loading will be discussed.

II. GROWTH FACTOR/CYTOKINE SIGNALING

The molecular events controlling the formation of RAS*-GTP follow the general pattern illustrated in Figure 3.1. Depending on the receptor system and the differentiation state of the cell there will be different tyrosine kinases, adaptors and exchange factors activated during the signaling process. The diversity of signaling commences with the complexity of each receptor system. Receptors can be conveniently considered as ligand-dependent protein kinases—some receptors encode

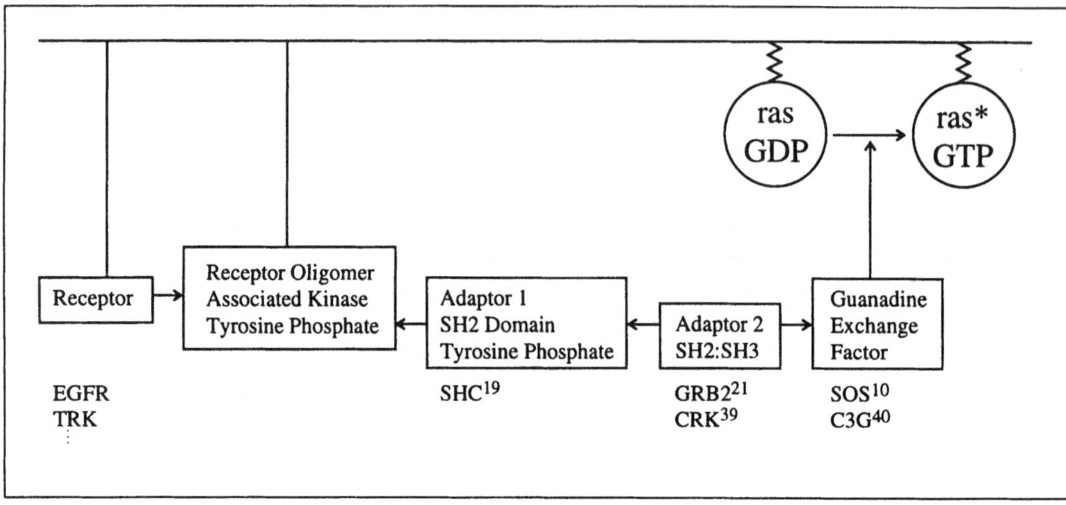

Fig. 3.1. Elements of the receptor activated pathway controlling the formation of ras-GTP.*

the kinase domain, other receptors bind an associated kinase. In most signaling systems several kinases associate with or are activated by the receptor. These kinases have diverse catalytic specificity and as a consequence, will activate multiple intracellular processes. Although several biochemical pathways have been associated with the activation of RAS*-GTP, the importance of these pathways is still being investigated. For example, the normal form of the EGF receptor kinase (ErbB1) clearly binds[41] and phosphorylates phospholipase C^{γ} (PLC^{γ});[42] however, a mitogenically active, but truncated form of the EGF receptor no longer binds PLC^{γ}.[42,43] Presumably the binding of PLC^{γ} is not a critical process for mitogenesis. There are many similar protein:protein interactions associated with the EGF receptor and it has been difficult to identify the events essential for mitogenesis. It is important to improve our understanding of the details of the receptor substrate (or adaptor) interactions if we are to identify the steps which are essential for mitogenesis.

EGF Receptor Family Signaling

Interestingly, the EGF receptor appears to exist in at least two distinct forms: an abundant low affinity form (K_D 1-10 nM)[44] and a scarce high affinity form (K_D~20 pM).[45] Mitogenic signaling correlates with the triggering of the high affinity receptors,[46,47] unfortunately we have an incomplete knowledge of the high affinity state. Several hypotheses have been developed to explain the high affinity state, including homodimers,[48] heterodimers[50] of the different EGF receptor family members (ErbB1 with ErbB2, ErbB3 or ErbB4) or a complex between the EGF receptor and an unknown affinity modulator.[51] Each of these complexes would be expected to stimulate a different range of intracellular signaling molecules, not all of which will be relevant to mitogenesis, but which may have a role in other cellular responses (e.g., morphogenesis or movement).

There are a number of growth factors capable of binding to the EGF receptor family;[52,53] however, we will not explore the signaling implications for the different ligands within this chapter. Suffice it to say that EGF,[54] TGF–α,[55] HB-EGF[56] and amphiregulin[57] might not interact identically with $(Erb1)_2$, (Erb1:Erb2), (Erb1:ErbB3 and ErbB1:ErbB4) and as a consequence, may be expected to induce distinct cellular responses. At present it still appears that an essential part of mitogenic signaling from the EGF receptor family involves the activation of RAS.

As mentioned earlier, the pathway connecting the EGF receptor and RAS was discovered by a combination of biochemical and genetic techniques. The initial mitogenic events depend on the activation of the receptor kinase.[46,58] A prominent substrate for the EGFR was recognized: (i) as the major tyrosine phosphorylated protein in liver after mice were treated with EGF[59] and (ii) as a transforming protein which contained an SH2 domain. These were the same protein called Shc.[19] When Shc is phosphorylated on tyrosine it is recognized by a second adaptor protein Grb2.[60] Since the SH2 domain of Shc binds to the EGF receptor it is easy to conceive how an initial signaling complex is established:

$$\text{EGFR-Y-(P)} \rightarrow \text{SH}_2\text{-Shc-Y(P)} \rightarrow \text{SH2-Grb2-SH3}$$

Grb2 is a small protein which contains one SH2 domain and two SH3 domains.[61] Genetic studies in yeast, worms and flies indicated that Grb2 activation was important for stimulating RAS.[62] The SH2 domain of Grb-2 is capable of binding to either the EGFR or Shc and the SH3 domain can bind to the guanine nucleotide releasing factor SOS.[63] The Grb-2:SOS complex is thought to translocate to the inner surface of the plasma membrane and stimulate the formation of an SOS→RAS-GDP complex.[64] Once the SOS-RAS-GDP complex has formed, the GDP is released and the free RAS is complexed by cytoplasmic GTP. The RAS*-GTP complex stimulates the kinase cascade headed by Raf-1 and leads to the activation of the MAP kinases.[66-68]

The EGFR can be phosphorylated at five tyrosine residues between Y992 and Y1173. Both Grb-2 and Shc can bind to several of these sites, creating the possibility of generating different signaling complexes from a single receptor. Depending on the extent of and distribution of these phosphorylation sites, many combinatorial complexes and signaling systems are possible.

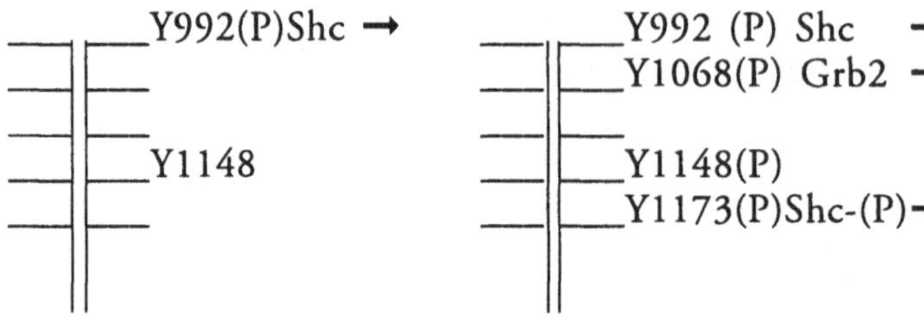

For example, other proteins such as PLC_γ, GAP and PI3 kinase can bind to these autophosphorylation sites to produce a complex of the general form

$$EGFR:(Shc)_n:(Grb\text{-}2)_m:X_l:Y_p:Z_q$$

where X, Y, Z are proteins containing SH2 domains capable of binding to phosphotyrosine residues and n-q are integers from 0 to 5.

Thus a single receptor can generate an enormous variety of signals. Of course, the final composition and potential of the EGFR signaling complexes will be controlled by the differentiation state of the target cell. Whilst Grb-2 can bind to the EGFR directly (via YP1068), most of the Grb-2 associated with the EGFR appears to be linked through Shc.[69] It should be remembered that the formation of the EGFR:Shc complex is not essential for Shc signaling. A C-terminally truncated EGFR kinase can still be activated by EGF, Shc is still phosphorylated and the Shc-Grb-2-SOS complex still mediates the activation of RAS.[70,71] As long as the Grb-2-SOS complex is activated, RAS will remain in its GTP form and continue to stimulate the Raf-1→MEK→MAPK cascade; however, some of the Ser/Thr kinases forming part of the cascade are also capable of phosphorylating SOS,[72] inducing the dissociation of the Grb2-SOS complex and allowing RAS*-GTP to return to its RAS*-GDP form.

The essential role for SOS in this RAS activation pathway is highlighted by the action of a dominant inhibitory analog of RAS, namely N-17-RAS.[26] N-17-RAS remains in the GDP form and thus sequesters most of the cellular SOS, reducing the ability of a mitogenic stimulus to load RAS with GTP. However, it must be noted that SOS is not the only guanine releasing factor (GRF).[22] There is a mammalian homologue of a yeast GDP releasing factor $CDC25^{Mm}$.[24,73] $CDC25^{Mm}$ (also called RAS-GRF[23])is usually expressed in neuronal cells and may well be involved in signaling from the nerve growth factor (NGF) receptor.

PDGF Receptor Signaling

While most receptor tyrosine kinases stimulate the formation of RAS*-GTP, the array of proteins associated with each receptor kinase is quite distinct. For example, Src associates with activated ErbB2 rather than activated ErbB1.[74] Furthermore, the links between the receptor kinase and the Grb-2-SOS complex may also be receptor-specific. The platelet derived growth factor (PDGF) receptor is autophosphorylated on tyrosine in the presence of PDGF. Many cytoplasmic proteins associate with the phosphorylated receptor; however, Grb-2 does not bind directly to the PDGF receptor. Instead, the PDGFR binds an SH2-containing protein tyrosine phosphatase (Syp—now called SHP-2)[75] and Syp acts as an adaptor for Grb-2, thus creating the link to activate SOS and then RAS.[76]

FGF Receptor Signaling

Acidic fibroblast growth factor (aFGF) stimulates the differentiation of PC12 cells, and activation of the RAS pathway is an essential element of this process.[77] Mutational studies which replace Tyr766 of the FGFR with Phe766 indicate that PLCγ-associated phosphatidyl inositol hydrolysis is not required for the induction of the Shc:Grb-2 complex,[78] activation of SOS or the phosphorylation of GAP.[77] The mutant receptors appear to have a reduced (three-fold) ability to stimulate the Raf-1 and MAP kinases—suggesting that these might be influenced by both RAS-dependent and PLCγ-dependent activation pathways. Interestingly, the Tyr766→Phe substitution of the FGF receptor still stimulates DNA synthesis in fibroblasts and differentiation in PC12 cells with an efficiency similar to that of the wild-type FGF receptor. RAS*-GTP accumulation in PC12 cells occurs after FGF stimulation, but as in many other cells, the three-fold increase in RAS*-GTP levels returns to basal levels after 60 minutes. In contrast, Nerve Growth Factor (NGF) induces a four-fold increase of RAS*-GTP levels in PC12 cells and this increase persists for at least 16 hours. In PC-12 cells, negative dominant N17-RAS blocks both the NGF-induced increase in RAS*-GTP and cell differentiation.

NGF and Neurotrophin Signaling

NGF signals via a set of tyrosine kinase receptors (TRKs).[79,80] After NGF activation of the TRK receptors in PC-12 cells, Shc is phosphorylated and binds to the SH2 domain of Crk-1.[81] Crk-1 also has an SH3 domain,[82] so it can act as an adaptor/activator of both SOS and RAS/Rap guanine releasing factor (C3G).[83,84] Interestingly, Grb-2 does not bind to the TRKA receptor and it appears likely that Crk-1 or Crk-2 acts as the adaptor for the GEF.[85] Overexpression of Crk-1 or Crk-2 does not alter the basal levels of RAS*-GTP; however, overexpression of this adapter amplifies the response to NGF and RAS*-GTP levels are elevated. Point mutations in either the SH2 or SH3 domain of Crk-1 inhibit NGF-induced activation of RAS.[81] Interestingly, the transforming gene of the avian sarcoma virus CT10[86] encodes a p47Gag-Crk fusion protein (v-Crk).[87] In PC12 cells, v-Crk not only enhances differentiation in response to NGF, but it also allows the EGF to induce differentiation (EGF does not induce the differentiation of the parental strain of PC-12 cells used for these experiments). Both EGF and NGF can stimulate the formation of a Shc/Grb2 complex in PC12 cells, but differentiation may also require the activation of the Crk/C3G complex.[88,89] It should be noted that Shc and Crk bind by different mechanisms to phosphotyrosine residues on TRKA—Crk via an SH2 domain, Shc via a distinct phosphotyrosine binding domain.[90] There is an indication that v-Crk binds directly to activated forms of the EGFR or TRKA receptors.[91] As distinct from other transforming genes, e.g., v-Src or v-RAS, v-Crk still requires an initial activation of a tyrosine kinase cascade (e.g., by EGF or NGF) before it can stimulate differentiation or transformation.[87] It is likely that the CT10 virus not only produces v-Crk but also induces an autocrine growth factor which constitutively activates an appropriate tyrosine kinase receptor.

NGF stimulates neuronal differentiation via RAS*-GTP.[92,93] NGF induces the activation of RAS both in cell lines such as PC12 cells and in primary embryonal sensory cells.[94,95] By contrast, brain-neurotrophic factor (BDNF) and neutrotrophin-3 (NT-3) do not activate RAS in sensory neurones. It appears that both the p75LNGF receptor and the p140TRK receptor need to be activated before the concentration of RAS*-GTP is increased and neuronal differentiation is induced.[94] It should noted that while the loading of RAS with GTP appears to be required for NGF-induced neurite formation, it is not sufficient for inducing neurite outgrowth.[96] A mutant form of the TRK-NGF receptor (missing the small segment Lys.Phe.Gly from the juxta-membrane region) does not mediate NGF-induced neurite outgrowth. This mutant receptor does mediate NGF stimulation of Shc, PLCγ and ERK tyrosine phosphorylation, thus suggesting the activation of RAS. There appears to be a RAS-independent pathway which is essential for NGF signaling of neurite outgrowth. Although an increase in RAS*-GTP is required for the NGF induction of neurite outgrowth, it is not required for the NGF-mediated survival of PC-12 cells in serum-free medium.[96] The prevention of apoptosis by NGF under these conditions appears to be mediated by PI-3 kinase.

Macrophage Colony Stimulating Factor Signaling

The production of myeloid and lymphoid cells requires the synergistic action of many growth factors/cytokines.[97-99] Most of these regulators act via complex receptor systems which activate associated tyrosine kinases; however, two hemopoietic regulators, macrophage colony stimulating factor (M-CSF, also called CSF-1) and stem cell factor (SCF), act via the receptor tyrosine kinases c-Fms[100] and c-Kit,[101] respectively.

M-CSF can regulate both the proliferation of early myeloid progenitor cells and the maturation of monocytic cells. As expected, stimulation of c-Fms by M-CSF results in autophosphorylation and the binding of PI3 kinase and Grb2 to the receptor.[102] However, in myeloid cells (FDC-P1) a protein known as p150 is also tyrosine phosphorylated in response to M-CSF.[103] The phosphorylated p150 can interact with the SH2 domain of Shc and Grb-2. Although Grb-2:SOS complex is preformed in FDC-P1 cells, it appears that activation of c-Fms induces the Grb-2-SOS complex to activate RAS. Since M-CSF can stimulate cell proliferation in both myeloid cells and fibroblasts, but p150 phosphorylation is only observed in stimulated myeloid cells. Thus p150 does not appear to be required for mitogenic signaling.

The effects of M-CSF on hemopoiesis appear to be quite complex: inducing self-renewal of early macrophage progenitors, synergizing with stem cell factor (SCF) to increase the frequency of self-renewal in multipotential precursors and controlling macrophage viability and function.[104] Many of the proliferative effects of M-CSF require the activation of the c-Myc gene and this process appears to be dependent on increased levels of RAS*-GTP.[105]

M-CSF induces the interaction of the Grb-2 protein with focal adhesion kinase p125-FAK.[106] This interaction is mediated via the binding of the Grb2-SH2 domain to the phosphotyrosine at Tyr925 on FAK. Whilst this complex can interact with SOS to activate RAS, the SH3 domain of Grb-2 also interacts with dynamin. Thus, it would not be surprising if M-CSF controlled secretory pathways associated with macrophage function.

Interleukin Signaling

Many nontyrosine kinase cytokine receptor systems[107-110] are known to mediate part of their signaling via the RAS system. Invariably, these receptors activate associated tyrosine kinases,[28,29] and some of these kinases mediate the activation of RAS.[109,110]

As with the receptor tyrosine kinases, multiple proteins are also activated by the nonkinase cytokine receptor systems; again it has been difficult to determine which events are critical for activating specific mitogenic or differentiative pathways. Activation of the interleukin-2 receptor stimulates T cell proliferation and an increase in the levels of RAS*-GTP.[111] After formation of the activated IL-2 receptor complex (an oligomer of the IL-2, IL-2α, β and γ receptor subunits) several tyrosine kinases are activated:[112] $p56^{lck}$ and $p59^{fyn}$ (Src family), Syk, JAK-1 and JAK-3. Both the serine rich and acidic regions of the IL-2β receptor are required for the activation of $p56^{lck}$, the activation of RAS and the induction of c-Fos;[113] however, only the serine rich region is required for activation of the c-Myc and Bcl-2 genes. It has been proposed that the activation of RAS is mediated by $p56^{lck}$ and, subsequently, the RAS-controlled MAP kinase cascade stimulates the induction of the Fos/Jun genes. p56^{lck} phosphorylates both the IL-2 Rβ chain and Shc, allowing the R→Shc→Grb-2:SOS→RAS-GTP link to be established directly.[114] However, it has been demonstrated that IL-2 can induce proliferation in cell lines which lack the Src family kinases normally expressed by T-lymphocytes.[115] Although Vav is present in T-cells and can cooperate with RAS* to transform cells, Vav is not a RAS GDP/GTP exchange factor.[116] Even in T-cells which lack p56^{lck}, Vav is tyrosine phosphorylated rapidly in response to IL-2. Experiments with inhibitors such as hErbimycin indicate that tyrosine kinases are involved in the IL-2 stimulation of RAS*-GTP loading.[117] Consequently, the pathway from the IL-2R to the activation of RAS is more appropriately described by

$$\text{IL-2R}\beta \rightarrow [\text{X}] \rightarrow \text{Shc(YP)} \rightarrow \text{Grb2} \rightarrow \text{SOS} \rightarrow \text{RAS-GTP}.$$

where X = unknown tyrosine kinase.

There is some evidence from hamster polyomavirus middle T-antigen studies that the Fyn kinase mediates IL-2 signaling,[118] but at present there is no direct evidence to definitely identify the tyrosine kinase (X) which links the IL-2R and RAS systems.

Several hemopoietins (e.g. interleukin-3 [IL-3], interleukin-5 [IL-5], granulocyte-macrophage colony stimulating factor [GM-CSF], stem cell factor [SCF]) stimulate tyrosine phosphorylation,[119] the phosphorylation of Shc and a decrease in the electrophoretic mobility of SOS.[120] Indeed, overexpression of Shc increases hemopoietin-mediated MAP kinase activation. However, it has been difficult to demonstrate that a Shc-Grb2-SOS complex forms on the plasma membrane as a result of hemopoietin stimulation.

Another pathway is activated by GM-CSF and IL-3: namely, the phosphatase Syp[122] (also called SHP-2),[75] is tyrosine phosphorylated and associates with Grb2. This interaction has also been observed in fibroblasts stimulated by PDGF,[76] but again the "physiological" significance of Syp as an adapter in the hemopoietin signaling systems awaits the results of further experiments.

Recent experiments have found that IL-5 activates the tyrosine kinase Lyn in eosinophils, and it has been suggested that Lyn is responsible for the increase in RAS-GTP.[123]

There is a set of cytokines, interleukin-6 (IL-6), leukemia inhibitory factor (LIF), oncostatin-M, ciliary neurotrophic factor (CNTF) and interleukin-11 (IL-11), which share a common receptor subunit, gp130.[124] IL-6 stimulation of PC12 cells induces neurite differentiation and a transient accumulation of RAS*-GTP.[125] Although neither gp130 nor the IL-6Rα chain is a kinase, the IL-6 dependent loading of RAS-GTP is dependent on the activity of a tyrosine kinase.[126] IL-6 induces tyrosine phosphorylation of gp130, JAK- and Shc. IL-6 and LIF induce the phosphorylation of gp130 and Hck (a Src family member).[127] Constitutive activation of Hck allows LIF-dependent embryonal stem cells to become less dependent on LIF, indicating that Hck may be involved in signaling from cytokines which activate the gp130 receptor.

III. GROWTH FACTOR CASCADES: AUTOCRINE STIMULATION

Although activated RAS plays an important role in cytokine signaling, there are clearly other pathways activated by the cytokine receptor and these pathways also determine the biological outcome of the signaling process. Whilst this chapter focuses on some of the biochemical pathways which activate RAS, it is important to remember that activated RAS also controls the expression of many growth factor/cytokine genes (see Fig. 3.2). Consequently, the action of an individual growth factor/cytokine may lead to the RAS*-dependent secretion of a number of growth factors/cytokines which can act on the cell in an autocrine fashion. These autocrine (RAS-mediated), "secondary" signals may amplify or sustain the activation of RAS, or even

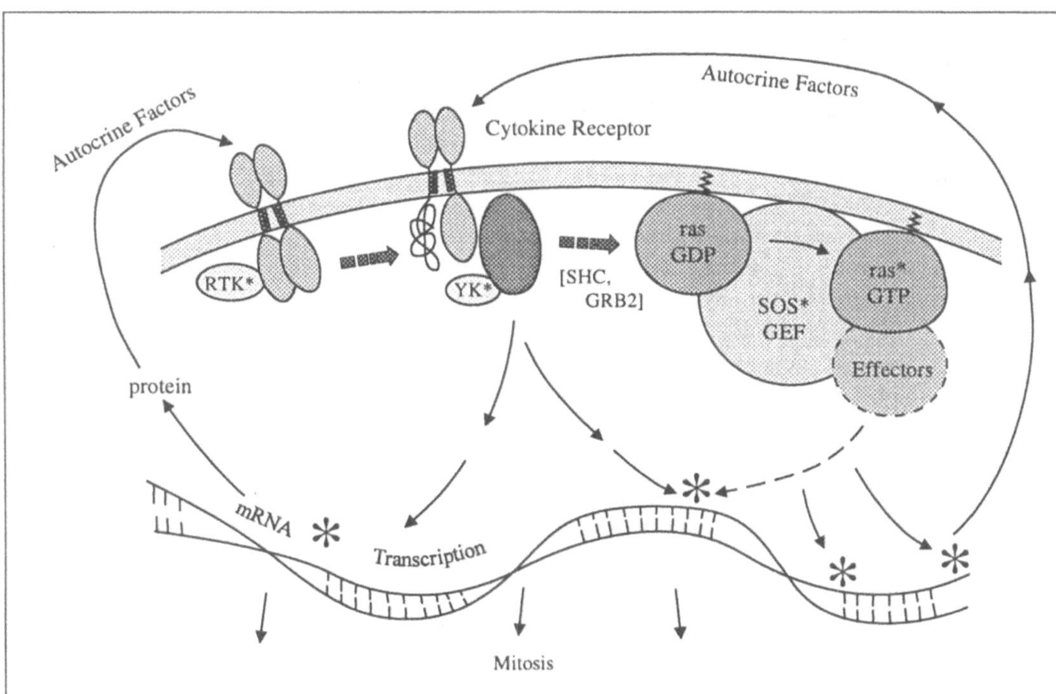

Fig. 3.2. Autocrine growth factor stimulation loops.

initiate other receptor-mediated events necessary for mitogenesis. One example of "induced synergy" is the action of TGF–α on fibroblast proliferation. TGF–α induces, in many RAS-dependent processes, including the expression of several EGF family members, amphiregulin and heparin-binding EGF (HB-EGF).[128] The mitogenic action of TGF–α appears to be dependent on this "autocrine" action of amphiregulin. Thus the ability of activated RAS to stimulate upstream processes (via the induced growth factor[s]) needs to be considered when analyzing the action of individual growth factors/cytokines or the effects of oncogenic RAS* on cell proliferation or differentiation.

The control of RAS*-GTP levels is critical for cell maturation, mobility and division. Activation of RAS is also likely to coordinate communication between cells in a given organ. In many cancers (perhaps as many as 30%) the control of RAS signaling is lost. This may be caused by constitutive growth factor production, mutations which activate receptors, constitutive activation of receptor-associated tyrosine kinases or the direct activation of RAS itself. It is reasonable to assume that if the constitutive RAS signaling can be blocked, it is likely that the cancer cells will undergo apoptotic death. The identification of agents capable of interfering with the receptor-RAS link should provide new opportunities for cancer treatment.

References

1. Harvey JJ. An unidentified virus which causes the rapid production of tumours in mice. Nature 1964; 204:1104-1105.
2. Shih C, Padhy LC, Murray M et al. Transforming genes of carcinomas and neuroblastomas introduced into mouse fibroblasts. Nature 1981; 290:261-264.
3. Krontiris TG, Cooper GM. Transforming activity of human tumor DNAs. Proc Natl Acad Sci USA 1981; 78:1181-1184.
4. McGrath JP, Capon DJ, Goeddel DV et al. Comparative biochemical properties of normal and activated human *RAS* p21 protein. Nature 1984;310:644-649.
5. Shih TY, Weeks MO, Young HA et al. Identification of a sarcoma virus coded phosphoprotein in nonproducer cells transformed by Kirsten or Harvey murine sarcoma virus. Virology 1979; 96:64-69
6. Shih TY, Papageorge AG, Stokes PE et al. Guanine nucleotide-binding and autophosphorylating activities associated with the p21src protein of Harvey murine sarcoma virus. Nature 1980; 287:686-691.
7. Gilman AG. G proteins and dual control of adenylate cyclase. Cell 1984; 36:577-579.
8. Fortini ME, Simon MA, Rubin GM. Signaling by the sevenless protein tyrosine kinase is mimicked by RAS-1 activation. Nature 1992; 355:559-561.
9. Aroian RV, Koga M, Mendel JA et al. The let-23 gene necessary for *Caenorhabditis elegans* vulval induction encodes a tyrosine kinase of the

EGF receptor subfamily. Nature 1990; 340:693-699.
10. Rogge RD, Karlovich CA, Banerjee U. Genetic dissection of a neurodevelopmental pathway: *Son of sevenless* functions downstream of sevenless and EGF receptor tyrosine kinases. Cell 1991; 64:39-48.
11. Chen J, Iyengar R. Suppression of RAS-induced transformation of NIH 3T3 cells by activated G alpha S. Science 1994; 263:1278-1281.
12. Ridley AJ, Paterson HF, Johnston CL et al. The small GTP-binding protein rac regulates growth factor-induced membrane ruffling. Cell 1992; 70:401-410.
13. De Larco JE, Todaro GJ. Growth factors from murine sarcoma virus-transformed cells. Proc Natl Acad Sci USA 1978; 75:4001-4005.
14. Burgering MT, Bos JL. Regulation of RAS-mediated signaling: more than one way to skin a cat. TIBS 1995; 20:18-22.
15. Maruta H, Burgess AW. Regulation of the RAS signaling network. BioEssays 1994; 16:489-495.
16. Sefton BM, Hunter T, Beemon K et al. Evidence that the phosphorylation of tyrosine is essential for cellular transformation by rous sarcoma virus. Cell 1980; 20:807-816.
17. Downward J, Parker P, Waterfield MD. Autophosphorylation sites on the epidermal growth factor receptor. Nature (London) 1984; 311:483-485.
18. Hu P, Margolis EY, Scolnik Y et al. Interaction of phosphatidylinositol 3-kinase-associated p85 with epidermal growth factor and platelet-derived growth factor receptors. Mol Cell Biol 1992; 12:981-990.
19. Pelicci G, Lanfrancone L, Grignani F et al. A novel transforming protein (SHC) with an SH2 domain is implicated in mitogenic signal transduction. Cell 1992; 70:93-104.
20. Pawson T, Schlessinger J. SH2 and SH3 domains. Curr Biol 1993; 13:3567-3576.
21. Lowenstein EJ, Daly RJ, Batzer AG et al. The SH2 and SH3 domain-containing protein Grb2 links receptor tyrosine kinases to RAS signaling. Cell 1992; 70:431-442.
22. Feig LA. Guanine-nucleotide exchange factors: a family of positive regulators of RAS and related GTPases. Curr Biol 1994; 6:204-211
23. Shou C, Farnsworth CL, Neel BG et al. Molecular cloning of cDNAs encoding a guanine-nucleotide releasing factor for RAS p21. Nature 1992; 358:351-354.
24. Martegani E, Vanoni M, Zippel R et al. Cloning by functional complementationn of a mouse cDNA encoding a homologue of CDC25, a *Saccharomyces cerevisiae* RAS activator. EMBO J 1992; 11:2151-2157.
25. Feramisco JR, Clark R, Wong G et al. Transient reversion of *RAS* oncogene-induced cell transformation by antibodies specific for amino acid 12 of *RAS* protein. Nature (London)1985; 314:639-641.
26. Feig LA, Cooper GM. Inhibition of NIH 3T3 cell proliferation by a mutant RAS protein with preferential affinity for GDP. Mol Cell Biol 1988 8:3235-3243.
27. Darnell JE Jr, Kerr IM, Stark GR. Jak-STAT pathways and transcrip-

tional activation in response to IFNs and other extracellular signaling proteins. Science 1994; 264:1415-1421.
28. Wilks AF, Harpur AG. Cytokine signal transduction and the JAK family of protein tyrosine kinases. BioEssays 1994; 16:313-320.
29. Ihle J. Cytokine receptor signaling. Nature 1995; 377:591-594.
30. Kazlauskas A, Cooper JA. Phosphorylation of the PDGF receptor β subunit creates a tight binding site for phosphatidyl 3-kinase. EMBO J 1990; 9:3279-3286.
31. Itoh T, Kaibuchi K, Matsuda T et al. A protein factor for RAS-dependent activation of MAP kinase through MAP kinase kinase. Proc Natl Acad Sci USA 1993; 90:975-979.
32. Trahey M, McCormick F. A cytoplasmic protein stimulates normal N-RAS GTPase, but does not affect oncogenic mutants. Science 1987; 238:542-545.
33. Manser E, Leung T, Salihuddin H et al. A non-receptor tyrosine kinase that inhibits the GTPase activity of p21cdc42. Nature 1993; 363:364-367.
34. DiBattiste D, Golubic M, Stacey D et al. Differences in the interaction of c-HaRAS-GMP-PNP with full-length neurofibromin (NF1) and GAP. Oncogene (1993); 8:637-643.
35. Roberts AB, Lamb LC, Newton DL et al. Transforming growth factors: Isolation of polypeptides from virally and chemically transformed cells by acid/ethanol extraction. Proc Natl Acad Sci USA 1980; 77:3494-3498.
36. Rayter SI, Woodrow M, Lucas SC et al. p21RAS mediates control of IL-2 gene promoter function in T cell activation. EMBO J 1992; 11:4549-4956.
37. Castelli C, Sensi M, Lupetti R et al. Expression of interleukin 1 alpha, interleukin-6, and tumor necrosis factor alpha genes in human melanoma clones is associated with that of mutated N-RAS oncogene. Cancer Res 1994; 54:4785-4790.
38. Kojima H, Hara K, Mineta-Kitajima R et al. Isolation of a subclonal cell line of PC12 transfected with dexamethasome-regulated RAS oncogene: morphological differentiation, biochemical properties, and tumorigenicity. J Biochem Tokyo 1993; 114:194-202.
39. Mayer BJ, Hamaguchi M, Hanafusa H. A novel viral oncogene with structural similarity to phospholipase C. Nature 1988; 332:272-275.
40. Tanaka S, Morishita T, Hashimoto Y et al. C3G, a guanine nucleotide-releasing protein expressed ubiquitously, binds to the Src homology 3 domains of CRK and GRB2/ASH proteins. Proc Natl Acad Sci USA 1994; 91:3443-3447.
41. Hernandez-Sotomayor SM, Carpenter G. Non-catalytic activation of phospholipase C-gamma 1 in vitro by epidermal growth factor receptor. Biochem J 1993; 293:507-511.
42. Posner I, Levitzki A. Kinetics of phosphorylation of the SH2-containing domain of phospholipase C gamma 1 by the epidermal growth factor receptor. FEBS Lett 1994; 107:109-116.
43. Gergel JR, McNamara DJ, Dobrusin EM et al. Identification of amino

acids in the N-terminal SH2 domain of phospholipase Cγ1 important in the interaction with epidermal growth factor receptor. Biochem 1994; 33:14671-14678.
44. Kawamoto T, Sato JD, Le A et al. Growth stimulation of A431 cells by epidermal growth factor: Identification of high-affinity receptors for epidermal growth factor by an anti-receptor monoclonal antibody. PNAS 1983; 80:1337-1341.
45. Rees AR, Gregoriou M, Johnson P et al. High affinity epidermal growth factor receptors on the surface of A431 cells have restricted lateral diffusion. EMBO J 1984; 3:1843-1847.
46. Fowler KJ, Walker F, Alexander W et al. A mutation in the epidermal growth factor receptor in waved-2 mice has a profound effect on receptor biochemistry that results in impaired lactation. Proc Natl Acad Sci USA 1995; 92:1465-1469.
47. Walker F, Nice E, Fabri L et al. Resistance to receptor-mediated degradation of a murine epidermal growth factor analogue (EGF-Val-47) potentiates its mitogenic activity. Biochem 1990; 29: 10635-10640.
48. Yarden Y, Schlessinger J. Self-phosphorylation of epidermal growth factor receptor: evidence for a model of intermolecular allosteric activation. Biochem 1987; 26:1434-1442.
49. Yarden Y, Schlessinger J. Epidermal growth factor induces rapid, reversible aggregation of the purified epidermal growth factor receptor. Biochem 1987; 26:1443-1451.
50. Qian X, Decker SJ, Greene MI. $P185^{c\text{-}neu}$ and epidermal growth factor receptor associate into a structure composed of activated kinases. Proc Natl Acad Sci USA 1992; 89:1330-1334.
51. Walker F, Burgess AW. Reconstitution of the high affinity epidermal growth factor receptor on cell-free membranes after transmodulation by platelet derived growth factor. Mol Cell Biol 1991; 266:2746-2752.
52. Groenen LC, Nice EC, Burgess AW. Structure-function relationships for the EGF/TGFα family of mitogens. Growth Factors 1994; 11:235-257.
53. Karunagaran D, Tzahar E, Liu N et al. Neu differentiation factor inhibits EGF binding. J Biol Chem 1995; 270:9982-9990.
54. Savage CR Jr, Inagami T et al. Epidermal growth factor. Location of disulfide bonds. J Biol Chem 1972; 247:7669-7672.
55. Marquardt H, Hinkapillar MW et al. Transforming growth factors produced by retrovirus-transformed rodent fibroblasts and human melanoma cells:Amino acid sequence homology with epidermal growth factor. Proc Natl Acad Sci USA 1983; 80:4684-4688.
56. Higashiyama S, Abraham JA et al. A heparin-binding growth factor secreted by macrophage-like cells that is related to EGF. Science 1991; 251:936-939.
57. Shoyab M, Plowman GD et al. Structure and function of human amphiregulin: A member of the epidermal growth factor family. Science 1989; 243:1074-1076.
58. Selva E, Raden DL, Davis RJ. Mitogen-activated protein kinase stimula-

tion by a tyrosine kinase-negative epidermal growth factor receptor. J Biol Chem 1993; 268:2250-2254.
59. Ruff-Jamison S, McGlade J, Pawson T et al. Epidermal growth factor stimulates the tyrosine phosphorylation of SHC in the mouse. J Biol Chem 1993; 268:7610-7612.
60. Rozakis-Adcock M, McGlade J, Mbamalu G et al. Association of the Shc and Grb2/Sem5 SH2-containing proteins is implicated in activation of the RAS pathways by tyrosine kinases. Nature 1992; 360:689-671.
61. Maignan S, Guilloteau J-P, Fromage N et al. Crystal structure of the mammalian Grb2 adaptor. Science 1995; 268:291-293.
62. Johnson GL, Vaillancourt RR. Sequential protein kinase reactions controlling cell growth and differentiation. Curr Biol 1994; 6:230-238.
63. Koch CA, Anderson D, Moran M et al. SH2 and SH3 domains: elements that control interactions of cytoplasmic signaling proteins. Science 1991; 252:668-674.
64. Buday L, Downward J. Epidermal growth factor regulates p21RAS through the formation of a complex of receptor, Grb2 adaptor protein, and SOS nucleotide exchange factor. Cell 1993; 73:611-620.
65. Moodie SA, Willumsen BM, Weber MJ et al. Complexes of RAS.GTP with Raf-1 and mitogen-activated protein kinase kinase. Science 1993; 260:1658-1661.
66. Buscher D, Hipskind RA, Krautwald S et al. RAS-dependent and -independent pathways target the mitogen-activated protein kinase network in macrophages. Mol Cell Biol 1995; 15:466-475.
67. Wane PH, Viciana PR, Downward J. Direct interaction of RAS and the amino-terminal region of raf-1 in vitro. Nature 1993; 364:352-355.
68. Cowley S, Paterson H, Kemp P et al. Activation of MAP kinase kinase is necessary and sufficient for PC12 differentiation and for transformation of NIH 3T3 cells. Cell 1994;77:841-852.
69. Batzer AG, Rotin D et al. Hierarchy of binding sites for Grb2 and Shc on the epidermal growth factor receptor. Mol Cell Biol 1994; 14: 5192-5201.
70. Soler C, Alvarez CV, Beguinot L et al. Potent SHC tyrosine phosphorylation by epidermal growth factor at low receptor density or in the absence of receptor autophosphorylation sites. Oncogene 1994; 9:2207-2215.
71. Gotoh N, Muroya K, Hattori S et al. The SH2 domain of Shc suppresses EGF-induced mitogenesis in a dominant negative manner. Oncogene 1995; 11:2525-2533.
72. Waters DB, Holt KH, Ross SE et al. Desensitization of RAS activation by a feedback dissassociation of the SOS-Grb2 complex. J Biol Chem 1995; 270:20883-20886.
73. Burton J, Roberts D, Montaldi M et al. A mammalian guanine-nucleotide-releasing protein enhances function of yeast secretory protein Sec4. Nature 1993; 361:464-467.
74. Muthuswamy SK, Siegel PM, Dankort DL et al. Mammary tumors expressing the *neu* proto-oncogene possess elevated c-Src tyrosine kinase

activity. Mol Cell Biol 1994; 14:735-743.
75. Adachi M, Fischer EH, Ilhe J et al. Mammalian SH2-containing protein tyrosine phosphatases. Letter to Editor. Cell 1996; 85:15.
76. Feng G-S, Hui C-C, Pawson T. SH2-containing phosphotyrosine phosphatase as a target of protein-tyrosine kinases. Science 1993; 259: 1607-1611.
77. Nakafuku M, Satoh T et al. Differentiation factors, including nerve growth factor, fibroblast growth factor, and interleukin-6, induce an accumulation of an active RAS.GTP complex in rat pheochromocytoma PC12 cells. J Biol Chem 1992; 267:19448-19454.
78. Spivak-Kroizman T, Mohammadi M, Hu P et al. Point mutation in the fibroblast growth factor receptor eliminates phosphatidylinositol hydrolysis without affecting neuronal differentiation of PC12 cells. J Biol Chem 1994; 269:14419-14423.
79. Heumann R. Neurotrophin signaling. Curr Opin Neurobiol 1994; 4:668-679.
80. Barbacid M. Neurotrophic factors and their receptors. Curr Biol 1995; 7:148-155.
81. Matsuda M, Hashimoto Y et al. CRK protein binds to two guanine nucleotide-releasing proteins for the RAS family and modulates nerve growth factor-induced activation of RAS in PC12 cells. Mol Cell Biol 1994; 14:5495-5500.
82. Tanaka S, Hattori S, Kurata T et al. Both the SH2 and SH3 domains of human CRK protein are required for neuronal differentiation of PC12 cells. Mol Cell Biol 1993; 13:4409-4415.
83. Matsuda M, Hashimoto Y, Muroya K et al. CRK protein binds to two guanine nucleotide-releasing proteins for the RAS family and modulates nerve growth factor-induced activation of RAS in PC12 cells. Mol Cell Biol 1994; 14:5495-5500.
84. Cordon-Cado C, Tapley P, Jing S. The *TRK* tyrosine kinase mediates the mitogenic properties of nerve growth factor and neurotrophin.3. Cell 1991; 66:173-183.
85. Suen K-L, Bustelo XR, Pawson T et al. Molecular cloning of the mouse *Grb2* gene: Differential interaction of the Grb2 adaptor protein with epidermal growth factor and nerve growth factor receptors. Mol Cell Biol 1993; 13:5500-5512.
86. Mayer BJ, Hamaguchi M, Hanafusa H. A novel viral oncogene with structural similarity to phospholipase C. Nature (London) 1988; 332:272-275.
87. Hempstead BL, Birge RB, Fajardo JE et al. Expression of the v-Crk oncogene product in PC12 cells results in rapid differentiation by both nerve growth factor- and epidermal growth factor-dependent pathways. Mol Cell Biol 1994; 14:1964-1971.
88. Knudsen BS, Fellet SM, Hanafusa H. Four proline-rich sequences of the guanine-nucleotide exchange factor C3G bind with unique specificity to the first Src homology 3 domain of Crk. J Biol Chem 1994; 269: 32781-32787.

89. Klein R, Jing S, Nanduri V et al. The *TRK* proto-oncogene encodes a receptor for nerve growth factor. Cell 1991; 65:189-197.
90. Dikic I, Batzer AG, Blaikie P et al. Shc binding to nerve growth factor receptor is mediated by the phosphotyrosine interaction domain. J Biol Chem 1995; 270:15125-15129.
91. Saltiel AR, Decker SJ. Cellular mechanisms of signal transduction for neurotrophins. BioEssays 1994; 16:405-411.
92. Wood KW, Qi H, D'Arcangelo G et al. The cytoplasmic raf oncogene induces a neuronal phenotype in PC12 cells: a potential role for cellular raf kinases in neuronal growth factor signal transduction. Proc Natl Acad Sci USA 1993; 90:5016-5020.
93. Nakafuku M, Kaziro Y. Epidermal growth factor and transforming growth factor-alpha can induce neuronal differentiation of rat pheochromocytoma PC12 cells under particular culture conditions. FEBS Lett 1993; 315:227-232.
94. Ng NF, Shooter EM. Activation of p21 RAS by nerve growth factor in embryonic sensory neurons and PC12 cells. J Biol Chem 1993; 268:25329-25333.
95. BoRASio GD, Markus A, Wittinghofer A et al. Involvement of RAS p21 in neutotrophin-induced response of sensory, but not sympathetic neurons. J Cell Biol 1993; 121:665-672.
96. Peng X, Greene LA, Kaplan DR et al. Deletion of a conserved juxtamembrane sequence in TRK abolishes NGF-promoted neurogenesis. Neuron 1995; 15:395-406.
97. Nicola NA. An introduction to the cytokines. In: Nicola NA, ed. Guidebook to Cytokines and Their Receptors 1st ed. New York:Oxford University Press.
98. Burgess AW, Tran TT. Grocyt Database. http://www/ludwig.edu.au/www/grocyt.html
99. Simpson RJ, Hammacher A, Smith DK et al. Interleukin-6: structure-function relationships. Protein Science 1996; in press.
100. Roussel MF. Signal transducton by the macrophage-colony-stimulating factor receptor (CSF-1R). J Cell Sci Suppl 1994; 18:105-108.
101. Turner AM, Bennett LG, Lin NL et al. Identification and characterization of a soluble c-*kit* receptor produced by human hematopoietic cell lines. Blood 1995; 85:2052-2058.
102. Saleem A, Kharbanda S, Yuan ZM et al. Monocyte colony-stimulating factor stimulates binding of phosphatidylinositol 3-kinase to Grb2.Sos complexes in human monocytes. J Biol Chem 1995; 270:10380-10383.
103. Lioubin MN, Myles GM, Carlberg K et al. Shc, Grb2, Sos1, and a 150 kilodalton tyrosine-phosphorylated protein form complexes with Fms in hematopoietic cells. Mol Cell Biol 1994; 14:5682-5691.
104. Brugger W, Mocklin W, Heimfeld S et al. Ex vivo expansion of enriched peripheral blood CD34+ progenitor cells by stem cell factor, interleukin-1 beta (IL-1 beta), IL-6, IL-3, interferon-gamma, and erythropoietin. Blood 1993; 81:2579-2584.

105. Baker DA, Maher J, Roberts IA et al. Evidence that RAS and myc mediate the synergy between SCF or M-CSF and other haemopoietic growth factors. Leukemia 1994; 8:1970-1981.
106. Kharbanda S, Saleem A, Yuan Z et al. Stimulation of human monocytes with a macrophage colony-stimulating factor induces a Grb2-mediated association of the focal adhesion kinase pp125FAK and dynamin. Proc Natl Acad Sci USA 1995; 92:6132-6136.
107. Shuai K. Interferon-activated signal transduction to the nucleus. Curr Biol 1994; 6:253-259.
108. Velazquez L, Fellows M, Stark GR. A protein tyrosine kinase in the interferon α/β signaling pathway. Cell 1992; 70:313-322.
109. Kishimoto T, Taga T, Akira S. Cytokine signal transduction. Cell 1994; 76:253-262.
110. Miyajima A, Mui AL, Ogorochi T et al. Receptors for granulocyte-macrophage colony-stimulating factor, interleukin-3, and interleukin-5. Blood 1993; 82:1960-1974.
111. Graves JD, Downward J, Izquierdo-Pastor M et al. The growth factor IL-2 activates p21RAS proteins in normal human T lymphocytes. J Immunol 1992; 148:2417-1422.
112. Minami Y, Taniguchi T. IL-2 signaling: recruitment and activation of multiple protein tyrosine kinases by the components of the IL-2 receptor. Curr Biol 1995; 7:156-162.
113. Satoh T, Minami Y, Kono T et al. Interleukin-2-induced activation of RAS requires two domains of interleukin 2 receptor beta subunit, the essential region for growth stimulation and Lck-binding domain. J Biol Chem 1992; 267:25423-25427.
114. Kobayashi N, Kono T, Hatakeyama M et al. Functional coupling of the *src*-family protein tyrosine kinases $p59^{fyn}$ and $p53/56^{lyn}$ with the interleukin-2 receptor: implications for redundancy and pleiotropism in cytokine signal transduction. Proc Natl Acad Sci USA 1993; 90:4201-4205.
115. Mills GB, Schmandt R, Gibson S et al. Transmembrane signaling by the interleukin-2 receptor: progress and conundrums. Semin Immunol 1993; 5:345-364.
116. Bustelo XR, Suen K-L, Leftheris K et al. Vav cooperates with Ras to transform rodent fibroblasts but is not a RAS GDP/GTP exchange factor. Oncogene 1994; 9:2405-2413.
117. Izquierdo M, Cantrell DA. Protein tyrosine kinases couple the interleukin-2 receptor to p21RAS. Eur J Immunol 1993; 23:131-135.
118. Brizuela L, Ulug ET, Jones MA et al. Induction of interleukin-2 transcription by the hamster polyomavirus middle T antigen: a role for Fyn in T cell signal transduction. Eur J Immunol 1995; 25:385-393.
119. Welham MJ, Dechert U, Leslie KB et al. Interleukin (IL)-3 and granulocyte/macrophage colony-stimulating factor, but not IL-4, induce tyrosine phosphorylation, activation, and association of SHPTP2 with Grb2 and phosphatidylinositol 3'-kinase. J Biol Chem 1994; 269:23764-23768.
120. Welham MJ, Duronio V, Leslie KB et al. Multiple hemopoietins, with

the exception of interleukin-4, induce modification of Shc and mSos1, but not their translocation. J Biol Chem 1994; 269:21165-21176.

121. Lanfrancone I, Pelicci G, Brizzi MF et al. Overexpression of Shc proteins potentiates the proliferative response to the granulocyte-macrophage colony-stimulating factor and recruitment of Grb2/SoS and Grb2/p140 complexes to the beta receptor subunit. Oncogene 1995; 10:907-917.
122. Feng, G-S, Hui C-C, Pawson T. SH2-containing phosphotyrosine phosphatase as a target of protein-tyrosine kinases. Science 1993; 259: 11607-1611.
123. Alam R, Pazdrak K, Stafford S et al. The interleukin-5/receptor interaction activates Lyn and Jak2 tyrosine kinases and propagates signals via the RAS-Raf-1-MAP kinase and the Jak-STAT pathways in eosinophils. Int Arch Allergy Immunol 1995; 107:226-227.
124. Hirano T, Matsuda T, Nakajima K. Signal transduction through gp130 that is shared among the receptors for the interleukin 6 related cytokine subfamily. Stem Cells Dayt 1994; 12:262-277.
125. Nakafuku M, Satoh T, Kaziro Y. Differentiation factors, including nerve growth factor, fibroblast growth factor, and interleukin-6, induce an accumulation of an active RAS.GTP complex in rat pheochromocytoma PC12 cells. J Biol Chem 1992; 267:19448-19454.
126. Murakami M, Hibi M, Nakagawa N et al. IL-6-induced homodimerization of gp130 and associated activation of a tyrosine kinase. Science 1993; 260:1808-1810.
127. Ernst M, Gearing DP et al. Function and biochemical association of Hck with the LIF-IL-6 receptor signal transducing subunit gp130 in embryonic stem cells. EMBO J 1994; 13:1574-1584.
128. Barnard JA, Graves-Deal R, Pittelkow MR et al. Auto- and cross-induction within the mammalian epidermal growth factor-related peptide family. J Biol Chem 1994; 269:22817-22822.

CHAPTER 4

Prenylation of RAS and Inhibitors of Prenyltransferases

Isabel Sattler and Fuyuhiko Tamanoi

I. INTRODUCTION

The recent development of farnesyltransferase inhibitors which inhibit the growth of *RAS*-transformed cells demonstrates that membrane association of RAS and other monomeric G proteins such as Rho is important for the tumorigenicity of RAS.[1-3] In the early 1980s, it was established that RAS is C-terminally modified and that this modification is important for both its membrane association and tumorigenicity. This observation was followed by the elucidation of the steps involved in the C-terminal modification. The first step in this modification is the addition of a farnesyl group to RAS utilizing farnesyl pyrophosphate (FPP) as the prenyl donor. Characterization of the farnesylation led to the identification of the enzyme farnesyltransferase (FTase). This enzyme and two related enzymes, geranylgeranyltransferases (GGTases) I and II, form a family of enzymes called protein prenyltransferases. The structure and function of these enzymes are being elucidated, which could provide vital information concerning the active sites of the enzyme. Inhibitors of these enzymes have been identified by a variety of approaches, and the inhibitors have been utilized to inhibit the growth of *RAS* tumors. In this chapter we will focus on two topics: (i) the structure and function of protein prenyltransferases and (ii) inhibitors of prenyltransferases. In the discussion, we have included observations on prenyltransferases in nonmammalian organisms such as plants, *Drosophila* and yeast. In our presentation of inhibitors

Regulation of the RAS Signaling Network, edited by Hiroshi Maruta and Antony W. Burgess.

of prenyltransferases we have tried to expand the view beyond the handful of peptidomimetic compounds that have been discussed elsewhere.[4,5] For related aspects on protein prenylation we refer to several other review articles.[4,6-10]

II. C-TERMINAL MODIFICATION OF RAS PROTEINS

RAS proteins undergo a series of posttranslational modifications at their C-termini.[1] This is illustrated in Figure 4.1 which shows the modification of the yeast RAS2 protein.[11-13] The protein is synthesized in the cytosol, and the first methionine is removed early in the synthesis presumably cotranslationally. Then, a farnesyl group is attached via a thioether linkage to a cysteine four amino acids from the C-terminus. Three C-terminal amino acids are removed by a protease and carboxymethylation of the C-terminal cysteine occurs by the action of a carboxymethyl transferase. This results in the accumulation of an intermediate form of RAS2 that can be detected in the cytosol. The final step of the modification involves the addition of palmitic acid to a cysteine located just upstream of the farnesylated cysteine. The mature RAS2 protein with a farnesyl and a palmitoyl group attached is predominantly localized to the plasma membrane. A similar series of modifications occurs with mammalian H-RAS and N-RAS proteins.[14,15] In biophysical studies it was shown that the presence of two hydrophobic chains significantly strengthens the membrane anchorage of proteins.[16] However, in the case of K-RAS4B, no palmitic group is attached. Instead, a stretch of basic amino acids called the "polybasic region" which is present upstream of the farnesylated cysteine facilitates the membrane association.[17,18] Thus, there are two different routes that RAS can take to associate with the plasma membrane: (i) farnesylation followed by palmitoylation, and (ii) farnesylation and a polybasic region. It also appears that the polybasic region of K-RAS4B has an additional function for substrate recognition. In vitro studies demonstrate that K-RAS4B can be recognized by GGTase I, and that the polybasic region is required for this recognition.[19] Thus, the polybasic region may enable the protein to be recognized by both FTase and GGTase I (see sections III and VI).

The C-terminal modification is crucial for the function of RAS. Mutants of oncogenic RAS which do not undergo the C-terminal processing are not capable of transforming cells.[20] Also, mutants of the yeast RAS protein which are not modified cannot complement the growth of a *RAS* mutant.[21] In addition, mutations of yeast defective in the modification can suppress phenotypes due to activated RAS.[22,23] These findings demonstrate that RAS needs to be membrane associated to function. Recent studies suggest that the C-terminal modification is also important for RAS proteins in order to allow activation of their effector molecules. Posttranslational modification is important for the ability of RAS to activate the MAP kinase cascade.[24,25] Furthermore,

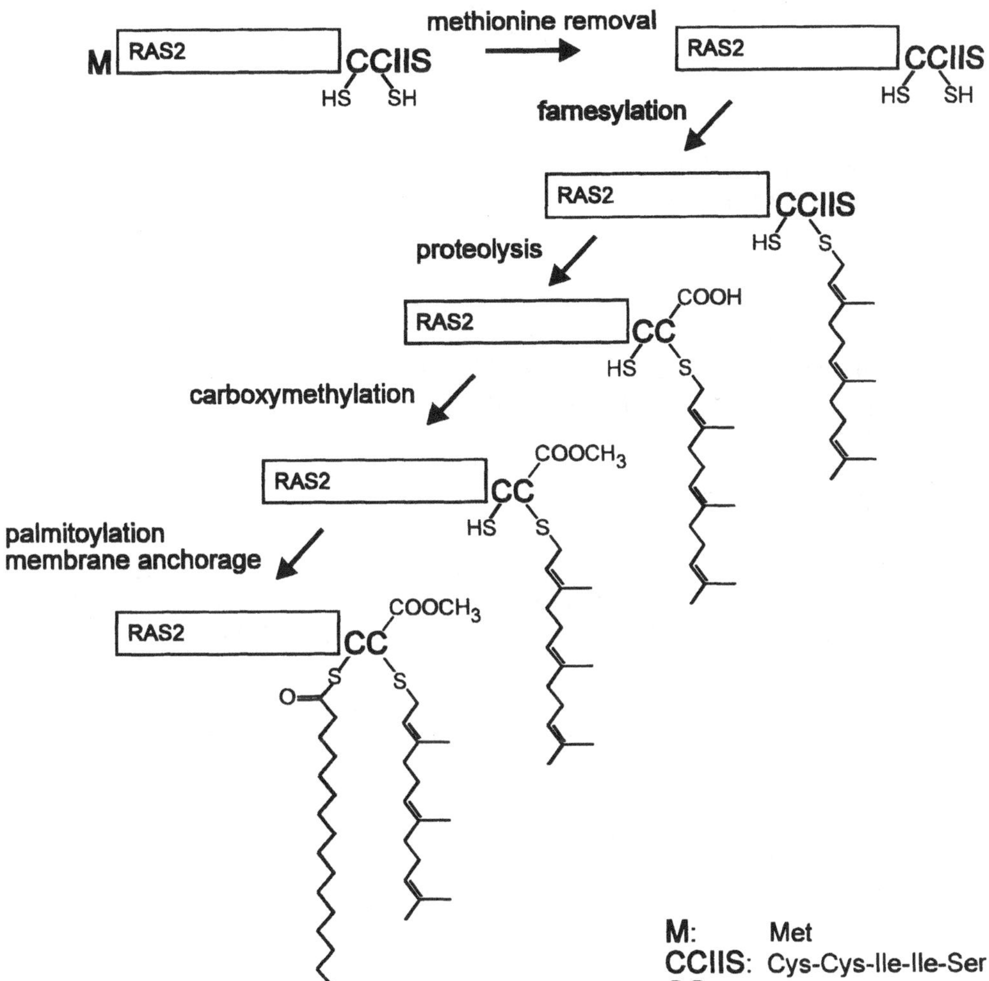

Fig. 4.1. Processing steps of RAS2 in Saccharomyces cerevisiae. *The figure shows a schematic depiction of RAS2, a protein with 322 amino acid residues, of which the C-terminal five amino acids are shown.*

posttranslationally modified yeast RAS2 protein is more efficient in activating its effector, adenylate cyclase.[26] Thus, the C-terminal modification may have additional significance in protein-protein interaction. Because activated RAS proteins do not transform cells if they lack the modification, RAS function can be downregulated by interfering with this series of modifications. This can be accomplished by blocking the farnesylation of RAS which is the first step in the C-terminal modification. It is also the step which is common to all RAS proteins and the only one which is absolutely required. A variety of farnesyltransferase inhibitors have been identified over the years and some of these have been shown to be effective in inhibiting the phenotypes of *RAS*-transformed cells (see section V).

III. FARNESYLTRANSFERASE AND GERANYLGERANYLTRANSFERASES

OVERVIEW

Protein prenyltransferases catalyze the addition of a prenyl group to the C-termini of a variety of members of the RAS-superfamily of G-proteins.[4] Three enzymes, farnesyltransferase (FTase), geranylgeranyltransferase type I (GGTase I) and geranylgeranyltransferase type II (GGTase II) have been identified. Figure 4.2 compares these enzymes. FTase catalyzes the addition of a farnesyl group from farnesylpyrophosphate (FPP) to proteins with the C-terminal CAAX sequence (A is an aliphatic amino acid). The C-terminal amino acid X in many cases is methionine, serine, cysteine, alanine or glutamine.[27-31] GGTase I catalyzes the addition of a geranylgeranyl group from geranylgeranylpyrophosphate (GGPP) to proteins ending with CAAL (the C-terminal amino acid usually is leucine or phenylalanine).[32-35] This enzyme is structurally and functionally very similar to FTase. Both FTase and GGTase I consist of two subunits, α and β. The α-subunit is shared

	FTase	**GGTase I**	**GGTase II**
subunit composition	α β	α β	α β REP
prenyl substrate	**FPP**	**GGPP**	**GGPP**
protein motif	**Cys-a-a-X** **X = Met, Ser, Gln, Ala, Cys**	**Cys-a-a-X** **X = Leu, Phe**	**Cys-Cys, Cys-X-Cys**
substrates	**Ras, nuclear lamins, STE18, transducin γ-subunit, Rhodopsin kinase**	**Rho, Rac, most trimeric G-protein γ-subunits**	**Rab**

Fig. 4.2. Characteristics of prenyltransferases. Subunit composition and substrate characteristics of FTase, GGTase I and GGTase II are summarized.

between FTase and GGTase I, while the β-subunit exhibits a significant homology between the two enzymes. GGTase II is another enzyme which utilizes GGPP. This enzyme catalyzes the addition of a geranylgeranyl group to two different types of substrate proteins, one type with a C-terminal CysCys and the other with a C-terminal CysXCys sequence.[28,36-40] Analyses of the geranylgeranylation of CysXCys ending proteins showed that both of these cysteine residues are geranylgeranylated.[41] GGTase II contains three subunits instead of two subunits. Two of them correspond to the α- and β-subunits of FTase and GGTase I. The third subunit is needed for the recognition of substrate proteins and is called the Rab escort protein (REP).[42,43] GGTase II recognition involves not only the C-terminal portion of the substrate protein but also other parts of the protein since proteins with only the CysCys or CysXCys sequence do not serve as substrates. The REP proteins provide this additional substrate recognition.

Prenyltransferases Have Been Identified in a Wide Range of Organisms

Prenyltransferases with similar structure and function have been identified in a wide range of organisms. Table 4.1 summarizes these findings in yeast, *Drosophila*, plant and mammalian cells. Mammalian genes encoding FTase subunits $\alpha_{F/GG}$ and β_F[44,45] as well as GGTase I subunits $\alpha_{F/GG}$ and β_{GG}[46] have been identified. The two subunits of GGTase II which correspond to the α and β subunits of FTase and GGTase I are termed α_{GGII} and β_{GGII}.[47] The third subunit of GGTase II is REP.[42,48] Two REP genes, REP-1 and REP-2, have been identified in mammalian cells.[49,50] The REP-1 and REP-2 gene products share a 75% identity. REP-1 is the product of the choroideremia (CHM) gene which is responsible for a retinal disease characterized by slow progressive peripheral retinal degeneration which leads to blindness.[49,50] Although the two REP proteins appear to have redundant functions, selective utilization of one of the REP proteins is observed in the case of one substrate, Rab27.[50]

Table 4.1. Genetic structure of prenyltransferases from different organisms

	FTase		GGTase I		GGTase II		
	β	α	β	α	β	α	REP
S. cerevisiae	DPR1(RAM1)	RAM2	CDC43(CAL1)	RAM2	BET2	BET4	MRS6(MSI4)
S. pombe	-	-	cwg2	-	ptb1	-	-
Drosophila	-	-	SR2-2	-	-	-	-
Plant	PsFTb	-	-	-	-	-	-
Mammalian	β_F	α_F	β_{GG}	α_{GG}	β_{GGII}	α_{GGII}	REP-1/REP-2

- = not identified

Similar enzymes have been purified from *Saccharomyces cerevisiae* and the genes encoding their subunits have been identified. DPR1/RAM1 and RAM2 correspond to the β and α subunits of mammalian FTase, respectively.[51,52] These genes were originally identified by mutations which suppress the phenotypes of yeast cells expressing activated RAS2.[22,23] It was shown that the processing of RAS is defective in these mutants, and that the mutants contained no detectable FTase activity. The *dpr1/ram1* and *ram2* mutants also exhibit defects in the maturation of one of the mating factors, a-factor, and display sterility specific to *MATa* cells. The *dpr1/ram1* mutation was also shown to suppress growth arrest due to the disruption of GPA1, the α-subunit of the yeast heterotrimeric G-protein.[53,54] The growth arrest is due to the unregulated activity of the βγ-subunits, which can be suppressed by blocking farnesylation of the γ-subunit. This observation formed the basis for a microbial screen to identify FTase inhibitors (see section V).[55,56] Two yeast genes, *CDC43/CAL1* and *BET2,* have been identified as genes with significant homologies to *DPR1/RAM1.*[57] The *CDC43/CAL1* and *BET2* genes were later found to encode the β-subunits of GGTase I and GGTase II, respectively.[34,35,40] *CDC43* is involved in cell polarity and bud emergence, a process that involves the function of G-proteins such as CDC42, RSR1 and RHO.[58] These proteins all end with the CAAL sequence. BET2 is involved in protein secretion, and mutations of this gene result in the loss of the membrane localization of YPT1 and SEC4 proteins.[40] *BET4,* the α-subunit of GGTase II was identified as a gene with homology to RAM2.[59] The yeast homologue of the *REP* gene is the *MRS6/MSI4* gene.[60,61]

In *Schizosaccharomyces pombe* the β-subunit of GGTase I was identified originally as a gene involved in cell wall synthesis.[62] This gene was suggested to encode a subunit of prenyltransferase because of its homology with prenyltransferase genes from other organisms. The assignment as to which prenyltransferase subunit this gene product corresponded was possible because the mutant was defective only in GGTase I activity. A G-protein that is involved in the synthesis of β-glucan, an important component of the cell wall, is presumably modified by GGTase I. A gene, *ptbI,* was identified by sequence homology as encoding the β-subunit of GGTase II.[63]

A recent surprise in the study of prenyltransferases was the identification of a *Drosophila* gene, *SR2-2.*[64] Mutations of this gene suppressed activated *RAS* phenotypes in *Drosophila* eye development, a process that involves *Drosophila* RAS1. The gene was cloned and found to share a significant homology with the β-subunit of GGTases I from other organisms. This, together with the fact that *Drosophila* RAS1 ends with a CAAL sequence, strongly suggests that GGTase I is involved in the prenylation of this RAS in *Drosophila.*

A gene from the garden pea (*Pisum sativum L.*) corresponding to the β-subunit of FTase has been isolated.[65] This gene, called *PsFTb,* was identified from a cDNA library constructed from apical bud RNA. An interesting observation about this gene is that its transcript levels within apical buds decline during photoregulated leaf development in continuous white light. This suggests that the expression of *PsFTb* is developmentally regulated.

Physiological Functions of Prenyltransferases and Cross-Specificity

The identification of prenyltransferases from a variety of organisms described above led to the realization that the enzymes perform different physiological roles. This is summarized in Table 4.2. For example, *S. cerevisiae* FTase is involved in regulating the RAS signaling pathway by modifying the RAS1 and RAS2 proteins. It is also involved in the modification of STE18, the γ-subunit of the yeast heterotrimeric G-protein.[53] GGTase I of *S. cerevisiae*, on the other hand, is involved in cell polarity and bud emergence.[34] Its major substrates are the CDC42 and RHO1 proteins. Finally, *S. cerevisiae* GGTase II is involved in the protein secretion process, and YPT1 and SEC4 are its major substrates.[40] A differential activity of FTase and GGTase may be important for overall physiology of cells. Recently, it was shown that there is a differential pattern of farnesylation and geranylgeranylation in different regions of the bovine brain, as well as in different tissues.[66]

One striking observation is that GGTase I plays an important role in regulating the function of RAS1 in *Drosophila*.[64,67] In addition, mammalian K-RAS4B can also be modified by GGTase I.[19] This is very different from the situation in *S. cerevisiae* where FTase is involved in the modification of RAS, and GGTase I is involved in the regulation of cell polarity. These recent results clearly point to the idea that both FTase and GGTase I are important for the prenylation of RAS proteins, which has important implications for the design of prenyltransferase inhibitors (see section VI).

Even though FTase and GGTase I have distinct functional characteristics, cross-specificity of FTase and GGTase I has been observed. Yeast cells lacking the β-subunit of FTase (*dpr1/ram1* null) can still grow, raising the possibility that RAS can be modified by GGTase I.[68]

Table 4.2. Functional involvement of prenyltransferases as defined by genetic analyses

	FTase	GGTase I	GGTase II
S. cerevisiae	Ras signaling	cell polarity	protein secretion
S. pombe	-	cell wall synthesis	-
Drosophila	-	Ras signaling	-

- = unknown

On the other hand, a defect in GGTase I can be suppressed by overproducing its substrates, RHO1 and CDC42. This suppression is particularly efficient when the C-termini of these substrates are changed to the CAAX sequence.[69] In support of the similarity between FTase and GGTase I, we found that it is possible to obtain a mutant form of yeast FTase which is capable of utilizing the CAAL motif as a substrate.[70] This occurs by a single amino acid change (see section IV). Finally, cross-specificity of the two enzymes has been observed in vitro and in vivo. FTase can utilize a GGTase I protein substrate, albeit at a reduced efficiency.[28,71] Cross-specificity of the two enzymes with regard to the utilization of the prenyl substrate also exists.[28,71,72] Both enzymes show a strong binding capacity in vitro for their respective "unnatural" prenyl substrate which in the case of GGTase I even results in an efficient farnesyl transfer to RhoB.[72] Previously, RhoB was shown to be either farnesylated or geranylgeranylated in COS cells.[73]

Enzyme Structure and Mechanisms

Mechanisms of the action of prenyltransferases have been studied extensively. These studies involved cross-linking approaches, kinetic analyses and the use of substrate analogs. In this chapter, we will concentrate on FTase and GGTase I, since these are the enzymes which have relevance to the prenylation of RAS.

FTase has a strong affinity for the CAAX peptide. In fact, this property was used to purify the enzyme from rat brain by CAAX peptide affinity chromatography.[27] The peptides also inhibit FTase activity with K_D values in the submicromolar range.[74] Only the tetrapeptide sequence is necessary for inhibition,. and aromatic residues such as in CVFM completely prevent substrate activity.[74] This CAAX peptide inhibition formed the basis for the development of peptide based FTase inhibitors (see section V). FTase exhibits an even stronger affinity for its other substrate, farnesyl pyrophosphate. Incubation of the enzyme with FPP results in the formation of a stable complex that can be isolated by gel filtration.[75] No covalent bond is formed, since FPP can be dissociated from the complex. Similar observations have been made with GGTase I. GGTase I activity is inhibited by CAAL tetrapeptides, and a complex consisting of GGTase I and GGPP can be isolated by gel filtration.[33]

Cross-linking studies of FTase have been carried out to identify which subunit is important for substrate binding. First, a photoactivated farnesyl pyrophosphate derivative was cross-linked to the β-subunit, suggesting that the β-subunit is responsible for the binding of farnesyl pyrophosphate.[76,77] When RAS protein was cross-linked to FTase, it was also attached to the β-subunit.[75] Similarly, peptides corresponding to the C-terminal CAAX sequence can be cross-linked to the β-subunit.[78] A longer peptide, on the other hand, cross-links to both α– and β-subunits, suggesting that the CAAX binding site is at the interface between the two subunits.[78] Thus, the β-subunit is important for the binding of both protein and FPP substrates. Binding of the peptide substrate has also been studied by NMR spectroscopy. In one study a type I β turn of the CVFM motif in a bound heptapeptide was determined,[79] whereas the peptidomimetic inhibitor L-739,787 (a close analog of L-739,750, see section V) displays a slightly different type III β turn conformation.[80]

FTase is a zinc-metalloenzyme and contains one mole of zinc per mole of enzyme.[81,82] Removal of zinc requires prolonged dialysis against EDTA. The activity of the dialyzed FTase can be restored by the addition of zinc and magnesium. Recently, two distinct FTases that are distinguishable by their zinc requirement for substrate binding and activity have been identified in human lymphoma Daudi cells.[83] GGTase I also contains zinc which can be removed by dialysis.[32,84,85] However, unlike FTase, the activity of GGTase I can be recovered by the addition of zinc alone. For both FTase and GGTase I, it was found that zinc is required for protein substrate binding but not for FPP binding. In GGTase I zinc can be replaced by cadmium with retention of the enzymatic activity.[85]

Kinetic analyses of FTase have been carried out by isotope partitioning studies as well as by presteady-state analyses.[86-88] The conclusions somewhat differ among different studies, but point to a similar conclusion; a preferred pathway proceeds through the enzyme-FPP complex formation. The enzyme first binds FPP and the binary complex then reacts with the protein substrate. In order to define the chemical mechanism of the prenylation reaction, the kinetic behavior of fluorosubstituted analogs of farnesylpyrophosphate was studied.[89] Generally, the prenylation reaction can take place by an electrophilic S_N1-type mechanism or a displacement reaction that in an S_N2 type fashion would proceed through a ternary "transition complex". When hydrogen, fluoromethyl and trifluoromethyl derivatives of FPP were used for alkylation of the sulfhydryl moiety, a dramatic decrease in K_m and k_{cat} values indicated that the prenylation follows a type of electrophilic alkylation.

IV. MUTATIONAL ANALYSES OF PRENYLTRANSFERASES

Two different approaches of mutational analyses have been utilized in order to gain information about the residues which might be involved in the substrate binding and catalytic activity of FTase. The first approach is to identify mutations which alter substrate utilization. As outlined in Figure 4.3, the idea is to randomly mutagenize FTase and select mutants that have acquired the ability to utilize GGTase I substrates. Such mutants should provide us with information about residues involved in substrate recognition. The second approach is the identification and characterization of inactivating mutations. Yeast mutants defective in FTase activity provided valuable insights that later

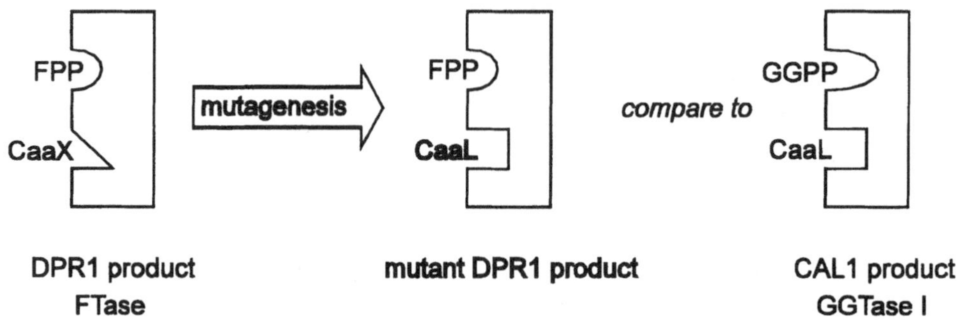

Fig. 4.3. Mutational conversion of peptide substrate specificity of FTase. Schematic depiction of a random mutagenic screen of the DPR1 product that complements a cal1 *mutation.*

were extended by in vitro mutagenesis. Also, mutants of FTase were devised which despite defects in their FTase activity, did display an intact interaction of the subunits.

FTase Mutants Exhibiting Altered Substrate Recognition

Utilizing the yeast system, we have obtained mutants of FTase which have acquired the ability to farnesylate GGTase I substrates.[70] The rationale behind such an approach is that FTase and GGTase I are structurally and functionally very similar. The two enzymes share the same α-subunit and their β-subunits display significant sequence homology. With respect to their protein substrates the only major difference between FTase and GGTase I concerns the C-terminal amino acid residue (CAAX versus CAAL). In our experiment *DPR1/RAM1* was randomly mutagenized and then screened in a yeast assay for mutants which were capable of suppressing a GGTase I defect (*cal1* mutant). This screen yielded three different mutants each containing a single amino acid change. One of these, mutant *S159N*, was examined in detail. This mutant suppressed the *cal1* temperature sensitive phenotype even on a single copy plasmid. It retained the ability to complement a *dpr1* mutation, suggesting that its ability to function as a FTase subunit was not drastically affected. Purification and characterization of the mutant enzyme showed that it has an approximately 20-fold increase in the affinity for GGTase I peptide substrates and 6-fold decrease in the affinity for its own peptide substrates. The mutant enzyme also exhibits increased sensitivity to a peptidomimetic FTase inhibitor, B581 (see section V). The mutation does not result in a relaxed substrate specificity for the prenyl transfer; it can not utilize GGPP. These results point to an altered recognition of protein substrates by this mutant enzyme. As mentioned above, two other mutations (Y362H, Y366N) exhibiting similar characteristics have been identified.[89a] A detailed study of these mutants may provide further information regarding the residues important for the recognition of protein substrates.

INACTIVATING MUTATIONS

Three mutations have been identified in the β-subunit of yeast prenyltransferases (*D209N, G259V* and *G328S*).[23,57,76] The corresponding amino acid changes (*D200N, G249V* and *G349S*) were introduced into mammalian FTase and the functional consequences have been examined.[76] The *D200N* and *G349S* mutations resulted in a decrease of protein substrate affinity without affecting the affinity for prenyl substrates. On the other hand, the *G249V* mutation resulted in a decrease in the affinity of both protein and prenyl substrates. None of these mutations affected subunit interaction, since mutant enzymes could be purified.

Mutations of the α-subunit have been identified by in vitro mutagenesis.[90] The deletion of 51 amino acids at the N-terminus of the α-subunit does not affect subunit interaction or FTase activity. However, deletion of 106 amino acids at the N-terminus abolishes both subunit interaction and FTase activity. Removal of 20 amino acids from the C-terminus abolishes FTase activity, whereas deletion of 5 amino acids reduces the activity to 50% of the control level. The substitution of a conserved lysine at residue 164 to asparagine abolishes FTase activity without affecting the subunit interaction. Mutations of four other conserved residues (*Y166F, R172E, N199D, W203H*) affect FTase activity to varying extents. The importance of residue 199 for FTase activity was further demonstrated by the analyses of the mutation *N199K*.[76] This mutation was identified in the yeast FTase mutant, *ram2* (corresponds to *N143K*). The *N199K* mutant has a significantly reduced affinity for the protein substrate whereas the affinity for the prenyl substrate is unaffected. This mutant has normal subunit interaction. Because the two inactivating mutations, *K164N* and *N199K*, did not affect the subunit interaction, it appears that the α-subunit plays some role in the catalytic activity.

V. INHIBITORS OF PRENYLATION

As farnesylation has been recognized as a possible target for controlling *RAS*, the search for inhibitors of FTase has become a very attractive approach for antitumor drug development.[2,3,5,91] The goal is to identify low molecular weight compounds that specifically and effectively inhibit FTase activity and which hold the potential for being developed into drugs with the necessary pharmacological properties. It has to be pointed out that recent findings suggest that geranylgeranylation might play a more important role in RAS modification than originally assumed. Until now, however, most efforts have been focused on identifying inhibitors that are specific for FTase over GGTase I.

Screening Approaches

Numerous inhibitors of FTase have been identified or synthesized. The various efforts by academic and industrial research groups in this area reflect the different routes that can be chosen in order to identify potent inhibitors of a given target system. In general, one can distinguish between approaches that either involve a targeted random screen or that are instead based on rational design. As the chance for hits in a random screen depends very much on a broad structural variety of the substance pool, natural products from microbial or plant sources are a premier source for identifying novel lead structures with this technique. "Classics" like the penicillins, but also recent examples like the immunosuppressive FK506 and the antitumor drug taxol, give powerful proof for the value of this method. We and others have successfully identified natural products from various sources that act as inhibitors of FTase, some of which exhibit potent activity and specificity. On the other hand, the rational design of enzyme inhibitors seeks to devise synthetic derivatives of the original substrates that can no longer perform their function. At the same time these substrate analogs should have retained or even improved binding to the respective site of the enzyme. This approach has yielded very effective and promising inhibitors for protein prenylation by employing FPP and more importantly the CAAX tetrapeptide as model compounds.

A feature that is crucial for the successful search for enzyme inhibitors is the monitoring assay that is applied. Several modifications of an in vitro assay for FTase activity have been established for a rapid and high throughput characterization of compounds.[92] Typically they assay the incorporation of a [^{3}H]-labeled prenylgroup into an acceptor protein that either is the recombinant original protein substrate or its C-terminus attached to some kind of carrier.[93-95] A fluorescence based assay with appropriately modified peptide substrates avoids the use of radioactive isotopes.[96,97] In contrast to an in vitro enzyme assay, an in vivo assay puts much higher demands on a substance in order to result in a positive hit. The need for membrane permeability and chemical stability in the cell ensures properties that can become crucial for the development of active compounds into drugs. In our laboratory an in vivo assay in a yeast system has been developed for the detection of FTase inhibitors.[56] The assay is based on a strain that carries a lethal disruption of the *GPA1* gene. FTase inhibitors rescue the cell cycle arrest caused by the lack of the α-subunit of the trimeric G-protein by blocking the farnesylation of the γ-subunit, thereby inactivating the constitutive signaling of the remaining βγ-complex.

Natural Products and Other Compounds Obtained by Random Screens

In ongoing screening programs by various research groups many natural products have been identified as potent inhibitors of FTase. Table 4.3 gives a summary of all compounds that have been identified by random screens and Figure 4.4 shows the chemical structures of these inhibitors. In the following we describe in detail those compounds and results that have been published since the last review in this area.[2] For additional information on pepticinnamin,[98] chaetomellic acid,[99-101] zaragozic acid[102] and 10'-desmethoxystreptonigrin[103] we refer to the literature.

Manumycin A and gliotoxin are the only natural products that have been shown to inhibit FTase in vivo.[55,56,104,105,105a] They were studied in two different yeast models, a *C. elegans* model system, and in a hepatoma cell line (see section VI). In addition to their inhibitory activity in the above described yeast *gpa1* mutant assay they were also shown to suppress RAS activation in yeast. Moreover, manumycin A was effective in a mouse tumor model with a required dosage that is comparable to those of peptidomimetic inhibitors (see section VI). However, some toxicity of manumycin has been noted in these experiments. Manumycin is one of the few natural product inhibitors that have been part of an SAR study with different derivatives.[55] In contrast to synthetic compounds, where derivatives are accessible as part of the synthetic concept, derivatives of natural products are usually not readily available. Manumycin, which has been isolated from two different *Streptomyces* strains, consists of a family of compounds termed manumycin A-D which vary in the branched side chain, as well as in the central cyclohexenone ring system.[106] The variations in the side chain result in some differences in the in vitro and in vivo inhibitory activity. A protocol from other studies on these compounds allows the oxidative removal of the triene chain with the five membered ring system.[107] The remaining part of the molecule shows comparable activity in the yeast *gpa1* assay, which indicates that the cyclohexenone ring system with the branched side chain is sufficient for FTase inhibition. Further studies on derivatives of manumycin are underway. As the structure with the unsaturated side chain already suggests, manumycin inhibition of FTase follows a FPP competitive mode. It displays some selectivity for FTase over GGTase I.

Actinoplanic acid, a metabolite from an *Actinoplanes* species is a potent and selective inhibitor of FTase.[108] Although it was identified as an FPP competitive inhibitor, it does not display any activity toward the FPP utilizing squalene synthase. Actinoplanic acid displays a novel complex structure of a 20-membered macrocyclic bislactone tetracarboxylic acid. Barceloneic acid A, which was isolated from a fungus species, has the advantage of being a FTase inhibitor of very low molecular weight.[109] Its inhibitory potential, however, is only modest.

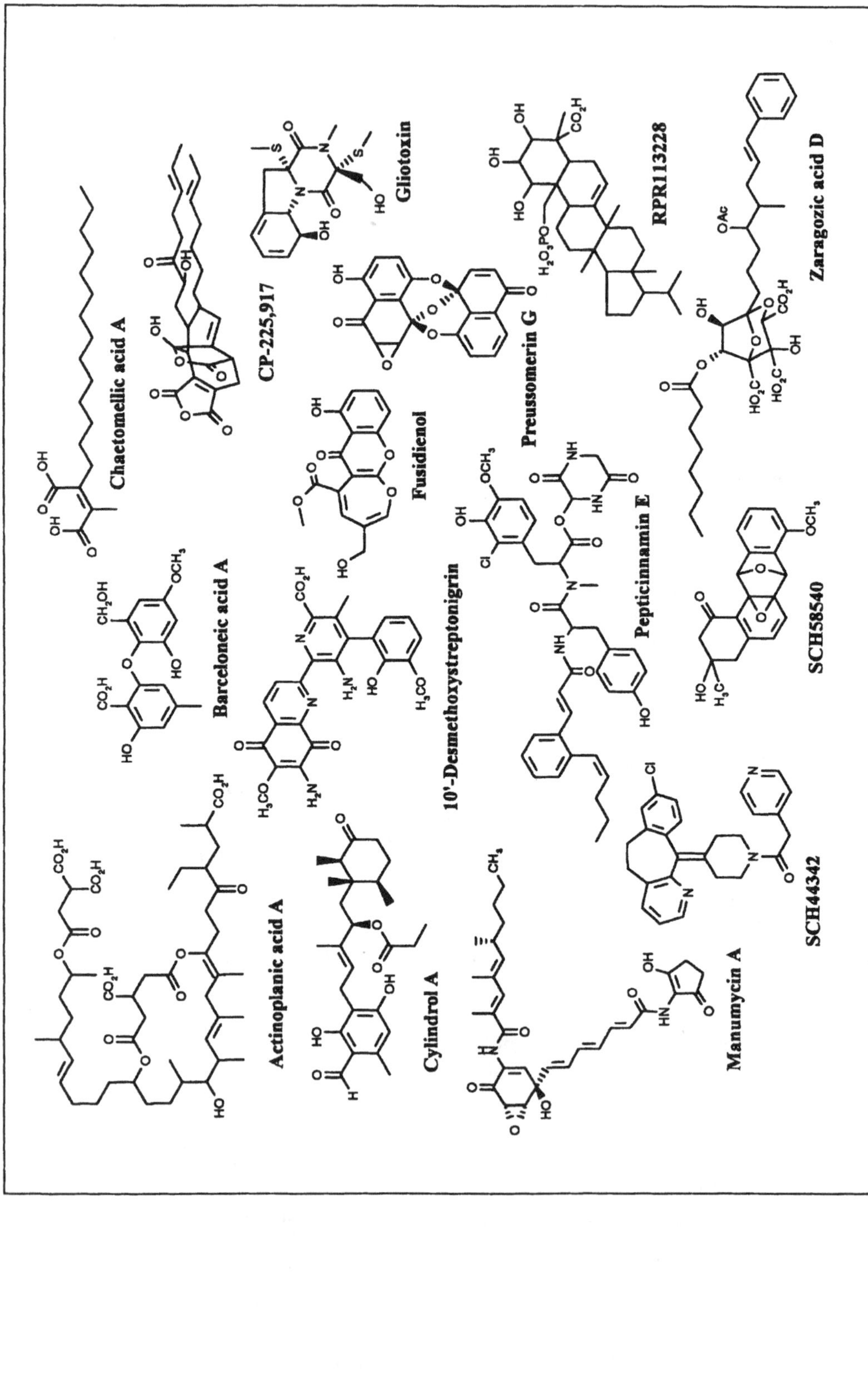

Fig. 4.4. Structures of natural products and other compounds obtained by random screen that inhibit FTase.

Table 4.3. Characteristics of FTase inhibitors obtained by random screen

Compound	IC_{50} (µM) FTase	GGTase I	GGTase II	Inhibition Mechanism	Remarks / Cellular Activity	Refs.
Actinoplanic Acid	0.23	>>1	ND	FPP competition K_i = 0.98 µM	no activity against squalene synthase, natural product	108
Barceloneic Acid A	40	ND	ND	ND	natural product	109
Chaetomellic acid A Chaetomellic Acid B	0.06 0.19	92 54	34 ND	FPP competition	inhibition of squalene synthase, natural product	99, 100, 101
CP-225,917 CP-263,114	6 6	ND ND	ND ND	ND ND	inhibition of squalene synthase, natural product	110
Cylindrol A	2.2	inactive	ND	no competition with either substrate	no activity against squalene synthase, natural product	111
10'-Desmethoxy-streptonigrin	21	ND	ND	ND	cytotoxic on human tumor cell lines, natural product	103
Fusidienol	0.3	inactive	ND	no competition with either substrate	no activity against squalene synthase, natural product	112
Gliotoxin	1.1	ND	ND	no competition with peptide	see Table 4.6, natural product	104 105
Manumycin A	5	180	ND	FPP competition	see Table 4.6, natural product	55, 56, 104, 105a
Pepticinnamin E	0.1	ND	ND	ND	natural product	98
Preussomerin G	1.2	20	ND	possibly irreversible	a family of structurally related natural products, of which derivatives G and D are the most active	113
RPR113228	2.1	59	ND	FPP competition, K_i = 0.4 µM	natural product	114
SCH58540	29	740	ND	ND	natural product	116
SCH44342	0.25	>114	ND	peptide competition, K_i = 0.24 µM	similar in vitro activity for H-Ras and K-Ras, inhibition of Ras processing in COS/H- at 3 µM (50%)	117
Zaragozic acid D	0.1	ND	ND	FPP competition	a family of structurally related natural products, of which D along with D_2 are the most potent FTase inhibitors; all of them are better inhibitors of squalene synthase	102

ND = not determined

A natural homologue in which the hydroxymethyl substituent is oxidized to an additional carboxylic acid group did not show any inhibitory activity despite only carrying a very minor structural change. The structurally unusual compound CP-225,917 and its analog CP-263,114 were reported to inhibit FTase and to some extent squalene synthase.[110] It is mentioned that the fungal metabolites act as reversible noncompetitive inhibitors of squalene synthase. In CP-263,114 the highly strained ring system is accompanied by another five membered ring which is formed by a lactol/ketal between the two hydroxy groups.

Another fungal metabolite, Cylindrol A, displays a medium activity with high selectivity towards FTase over GGTase I.[111] The novel bicyclic compound is noncompetitive with both enzyme substrates and is inactive against squalene synthase. Fusidienol, isolated from *Fusidium griseum*, is one of the inhibitors in this group with submicromolar activity.[112] Being inactive against GGTase I it also displays a very good selectivity. The mechanism of inhibition is unknown, and the compound is noncompetitive with respect to both substrates. Fusidienol represents a novel tricyclic oxygen-containing ring system.

Preussomerins and related compounds have recently been described from different fungal sources.[113] These compounds comprise two decalin type ring systems of different oxidation states that are connected to varying degrees by spiroketal bridges. Preussomerin G along with preussomerin D, in which the carbonyl next to the epoxide is reduced to a hydroxy function, as well as a synthetic analog with opened epoxide function, are all potent FTase inhibitors. Preussomerin G has a 18-fold selectivity for FTase over GGTase I. Other natural and synthetic analogs are less active or display no inhibition at all. Compared to preussomerin G they either lack one or both of the following features: the oxygen functionalities at the epoxide carrying ring system or the conjugated ketone of the "lower" ring system. This points towards a mode of action in which the conjugated ketone may serve as an acceptor for the nucleophilic sulfhydryl group of the cysteine residue.

RPR113228, which displays a steroid-type skeleton, was identified as FTase inhibitor with good activity and a 28-fold selectivity over GGTase I.[114] Mechanistic studies indicate that this fungal metabolite is competitive with respect to the prenyl substrate. It is inactive against squalene synthase. The compound carries a phosphate group and has a highly hydrophilic moiety at one end of the otherwise hydrophobic ring system. Similar structural features can be found in DHEAS (adrenal steroid dehydroepiandrosterone), a substance implicated in suppressing the occurrence of cancer.[115] DHEAS has been shown to have a modest FTase inhibition in vitro. Both steroid type compounds possess a polar region at ring A which is much more pronounced in RPR113228. The benzanthracene SCH58540 was isolated from a *Streptomyces* species.[116] It is a modest inhibitor of FTase that has 25-fold selectivity over GGTase I; however the inhibitory mechanism has not been described.

The tricyclic SCH44342 has been described recently as a compound with inhibitory characteristics of a peptidomimetic.[117] It was identified from a random screen of synthetic compounds. The structure is entirely nonpeptidic and does not offer obvious clues for its binding to the peptide recognition site. SCH44342 is of medium in vitro potency but is of particular interest because it has good cellular activity in inhibiting *RAS*-transformed phenotypes (see section VI).

Finally, the monoterpene limonene and its major metabolite perillyl alcohol are known as therapeutic agents against rat mammary cancer, and limonene is currently undergoing clinical testing.[118] These compounds have also been implicated in the inhibition of RAS processing in NIH3T3 and mammary epithelial cells. The in vitro potential of limonene and some of its oxidation products has recently been studied.[119] The terpene, as well as its alcohol and carboxylic acid metabolites, display only modest inhibition of FTase. The minor physiological metabolite perillic methyl ester, however, has some inhibitory activity, albeit little selectivity against GGTase I.

Comparing the different compounds that have been identified as FTase inhibitors by random screen, a striking diversity in structure and molecular sizes can be noted. In order to take the inhibitor development to the next stage, it would be critical to identify crucial structural features and use those for a rational molecular design approach. Since there are only fragmentary data available about the inhibitory mechanisms of these compounds it is difficult to rigorously deduce an optimal inhibitor. However, there are some common structural features which we would like to point out. Some of the inhibitors, namely zaragozic acids, chaetomellic acids, manumycins, actinoplanic acid and CP-225,917 possess branched or linear carbon chains as a common structural feature. Interestingly, all but one have been reported to act as FPP competitive inhibitors, thereby suggesting that they bind to FTase in the area of the FPP binding site. Indeed, for chaetomellic acid A it was shown by computer modeling that its carbon chain emulates the hydrophobic region of FPP and that the dicarboxylate mimics the electrostatic arrangement of the pyrophosphate.[101] The only other compound that has been described as a FPP competitor, RPR113228, shows an extensive hydrophobic portion that might act in a similar manner. Another structural feature contained in many of the described compounds is the abundance of polarizing oxygen substituents, especially carbonyls. These might be useful for the establishment of hydrogen bonds with the enzyme. It is also remarkable that many of the compounds have small ring systems. Generally, ring systems hold their substituents in a better defined, more rigid spatial arrangement than open chain compounds. One conclusion might be that these compounds bind to the enzyme at more than one position, which would be facilitated in a structurally well defined system as opposed to a flexible chain. However, further information about specific enzyme/inhibitor interactions as well as knowledge about the enzyme structure are needed to draw detailed conclusions.

FPP Analogs

At first sight FPP analogs seem to be the natural choice as substrate analog inhibitors of FTase.[120] In contrast to peptidomimetics they are of low molecular weight and do not possess critical peptidic moieties. Some problems have to be solved, though, in order to develop this class of compounds into potential drugs. One is the replacement of the highly charged allylic phosphate moiety with a biosteric unit that enhances membrane permeability and chemical stability. The other problem is to ensure selectivity for FTase over other FPP utilizing enzymes such as those involved in squalene synthesis or in the synthesis of ubiquinone or heme a. In the following, we describe those FPP analogs that show the most promising properties towards FTase inhibition (Table 4.4 and Fig. 4.5).

(α-hydroxyfarnesyl)phosphonic acid (**1**) was the first synthetic FPP analog reported to inhibit FTase in vitro.[100] It has a potent activity of an IC_{50} of 30 nM with a 1000-fold specificity for FTase over the two GGTases. It also inhibits squalene synthase with a submicromolar IC_{50} value. Compared to that, the structurally much more simple FPP analog farnesylamine (**2**) is a poor FTase inhibitor, although it has some cell activity.[121]

A set of FPP analogs were synthesized in which the very labile links of the pyrophosphate moiety and the attachment to the isoprenoid were individually replaced.[122,123] This study looked at FTase inhibitory

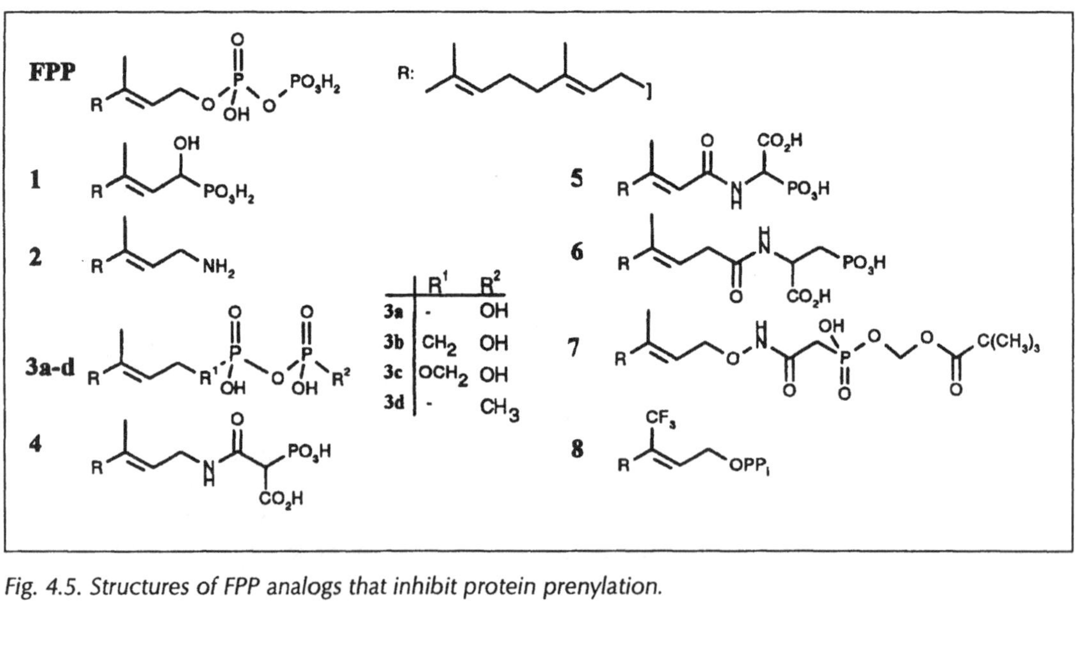

Fig. 4.5. Structures of FPP analogs that inhibit protein prenylation.

Table 4.4. Characteristics of FPP analogs inhibiting protein prenylation

Compound	IC_{50} [μM] FTase	GGTase I	GGTase II	Cellular RAS Processing	Remarks	Refs.
1	0.03	35.8	34.0	- inhibition of RAS processing in NIH3T3/H-RAS above 1 μM - no activity against Rat1/H-RAS-CVLL up to 100 μM	- cytotoxicity over 100 μM - squalene synthase IC_{50} = 0.63 μM	100
2	~22	~115	ND	- inhibition of proliferation of T24 NIH3T3/H-RAS (PAP2/H-RAS) at 7.5 μM	- cell viability up to 50 μM	121
3a	0.34	ND	ND	- inhibition of RAS processing in Rat1/H-RAS13 above 1 μM with modest specificity over general protein prenylation	- no general toxicity: no effect on cholesterol biosynthesis and overall protein biosynthesis - squalene synthase IC_{50} = 129 μM	121,122
3b	0.93	ND	ND	ND	- squalene synthase IC_{50} = 518 μM	121,122
3c	0.5	ND	ND	ND		121,122
3d	8.9	ND	ND	ND	- no inhibition of squalene synthase	121,122
4	0.075	ND	ND	ND	- IC_{50} > 200 μM against mevalonate kinase that is feedback inhibited by FPP - squalene synthase IC_{50} = 516 μM	92,120
5	0.25	ND	ND	ND		120
6 (L) **6** (D)	0.05 1.7	ND	ND	ND		120
7 free acid	2.6 0.075	24.5 ND	24.2 ND	- Inhibition of RAS processing in NIH3T3/K-RAS 4B (100% at 100 μM) with modest specificity over general prenylation - reversion of morphology of NIH3T3/H-RAS at 100 μM	- no cytotoxicity - squalene synthase IC_{50} = 95 μM	92,120
8	ND	ND	ND	ND	- K_i = 11 nM, 770-fold less reactive than FPP	89

ND: not determined

activity as well as squalene synthase activity, thereby exploring whether specific inhibition of either enzyme is possible. The most potent inhibitor of squalene synthase, a derivative of **3d** with methylene group as linker in the pyrophosphate unit, turned out to be a weak inhibitor of FTase. This suggests that the oxygen atom in this position is of particular importance for binding to FTase. In addition, results with compounds **3a** and **3c** strongly support the possibility of selectively inhibiting either one of the two FPP utilizing enzymes. The negatively charged oxygen atoms at the terminal phosphate group seem to be another important structural feature necessary for binding to the enzyme. Substitution with a methyl group results in a dramatic loss of activity (**3d** as opposed to **3a** and the phosphonate analog of **3b**). This corresponds to a model of the FPP binding pocket derived from research on squalene synthase inhibitors (as cholesterol lowering agents) in which the highly charged pyrophosphate moiety is stabilized by complexation with a magnesium ion and by electrostatic interaction with a positively charged amino acid residue of the protein. This model was also used to develop FPP analog **4** in which an amide linker unit is introduced and a carboxylic acid residue is used to approximate the arrangement of charges in the pyrophosphate moiety.[92] The potency and selectivity of **4** prompted a systematic SAR study involving modifications of the farnesyl group, the amide linker and the pyrophosphate surrogate.[120] Compounds **5**, **6** and the free acid of **7** represent the most potent inhibitors which originated from this study which was comprised of more than forty FPP analogs. It turns out that the carboxylic and the phosphonic acid groups in the hydrophilic region are essential for activity. Any changes in the β-carboxylic phosphonate moiety result in 50- to 2600-fold loss in inhibitory activity. It is also striking that the farnesyl binding pocket apparently has very stringent structural requirements for accommodating the hydrophobic portion. For example, the insertion of a single methylene unit next to the amide nitrogen resulted in a 200-fold loss of activity. Changes in the amide linker as exemplified in **5** and **6** seem to be much better tolerated. It is noteworthy that the two stereoisomers of **6** differ in their activity 34-fold. Other enantiomers in this series of derivatives show much smaller differences in their inhibitory activity. The parent compound of **7** shows a similar in vitro potency as **4** and **6** but has the advantage of only having one highly charged acid group.[92] Due to permeability problems it still performs poorly in cell assays. Therefore the pivaloyloxymethylester **7** was devised as a prodrug that releases the active compound upon esterase hydrolysis. Compound **8** was not developed as inhibitor of FTase. It was part of a kinetic study of FTase in order to gain insight into the reaction mechanism of protein farnesylation (see section III).[89] It has 770-fold less reactivity than FPP itself.

Peptidomimetics

Much effort for the development of FTase inhibitors has been made in the area of peptidomimetic compounds which are modeled after the C-terminus of the original substrates.[5] The compounds are summarized in Table 4.5 together with their in vitro activities and results about in vivo RAS processing. Figure 4.6 depicts their chemical structures. The initial effort in the design of peptidomimetic inhibitors focused on defining the structural features of the CA_1A_2X motif which are required for recognition and binding by the enzyme as well as those which prevent the prenylation reaction.[124,125] The prototype of a competitive tetrapeptide inhibitor is CVFM (**10**). Cysteine is required as the N-terminal residue, and its distance from the C-terminus is crucial for recognition by FTase. As discussed earlier, the C-terminal residue governs the recognition by FTase or GGTase I. For FTase inhibition this residue is preferably methionine or serine, whereas leucine enhances the recognition by GGTase I. Incorporation of different amino acids indicated a variability in the A_1A_2 portion, which however has to be hydrophobic.[29,74] Phenylalanine or other aromatic residues at the A_2-position prevent farnesylation. Another structural requirement for preventing farnesylation of the peptide is an unblocked amino function at the N-terminus. Retention of the negative charge at the carboxylate of the C-terminus is required for good inhibitory activity.[124] Other tetrapeptides that have been used as templates for inhibitor development are CIFM, CVIM and CIIM.

A most critical problem for drug development from peptidic compounds is their chemical properties that result in poor solubility, poor membrane permeability and susceptibility to proteolysis. Different modifications of the internal peptidomimetic moiety of the CAAX analog have been used in order to overcome the stability problems. For in vivo applications, prodrug strategies are necessary for temporary masking of the C-terminal carboxylate in order to allow membrane passage. This may be accomplished by lactone formation of a homoserine residue like in L-731,734 (**11b**) or more commonly by esterification with methanol or isopropanol.

In the following summary of peptidomimetic inhibitors, we have not only included those compounds which are now in the spotlight because of detailed biological studies, like B581, BZA-2B, L-739,749 or FTI-276, but also examples of compounds which have been developed along the way in order to illustrate the structural variety of inhibitors.

Various studies have employed backbone modifications in order to render the peptide structure more stable.[125] The first two peptide bonds from the C-terminus can be reduced to ethylamine moieties or replaced by a carbon tether or an ether linkage without a major effect on their in vitro activities (**11-14**). L-731,735,[125,127] B581[128] and L-739,750[129] are the most prominent results of this strategy. Compounds

Table 4.5. Characteristics of peptidomimetics and a bisubstrate analog inhibiting protein prenylation

	IC_{50} [nM]		Inhibition of RAS Processing:		
Compound	**FTase**	**GGTase I**	**- Concentration**	**- % Inhibition and Cell System**	**Refs.**
CVFM (**10**)	57	11000			74, 124-126
CIFM	27	300000			"
CVIM	165	49000			"
CIIM	150	ND			"
L-731,735 (**11a**)	18	>100000	~100 µM	50% in NIH3T3/H-RAS by L-731,734 (**11b**)	126, 127, 129
12a	29	15000			130
12b	470	ND			130
12c	3.5	3300	~0.1 µM ~2.5 µM	50% in NIH3T3/H-RAS by **12e** 50% in NIH3T3/H-RAS by **12f**	130
12d	20	ND			130
B581 (**13**)	21	790	~50 µM	50% in NIH3T3/H-RAS	128, 143
B956	11	ND	0.5 µM 25 µM	50% in NIH3T3/H-RAS (zH1) 50% in NIH3T3/K-RAS (DK1)	131
L-739,750 (**14a**)	1.8	3000	0.1-1 µM	50% in NIH3T3//H-RAS by L-739,749 (**14b**)	129
15	29000	ND			132
C-AMBA-Met (**16**)	60	ND	200 µM	Balbc3T3/H-RAS	134, 135

FTI-232 (**17a**)	150	1500	200 µM	100% in NIH3T3/H-RAS by FTI-244 (**17b**)	67, 136
FTI-249 (**17c**)	300	4400	200 µM	100% in Balbc3T3/H-RAS by FTI-254 (**17d**)	67, 137
FTI-279 (**17e**)	ND	100			67, 137
18a	< 2000	~95000			133
18b	20	4100			133
18c	9	90			133
18d	8	5100			133
FTI-276 (**19a**)	0.5 (K-RAS)	50	0.1 µM, 10 µM, ND	50% in NIH3T3/H-RAS, 50% in NIH3T3/K-RAS4B, NIH3T3 (endogenous RAS) by FTI-277 (**19b**)	67, 137
GGTI-287 (**19c**)	25	5	2 µM, >30 µM	inNIH3T3/K-RAS4B, in NIH3T3/H-RAS by GGTI-286 (**19d**)	67, 137
FTI-265 (**20**)	150	100000	50 µM	in Balbc3T3/H-RAS	67, 136
BZA-2B (**21a**)	0.9 (K-RAS)		10-50 µM	in Met18b-2 (CHO), Rat1/H-RAS by BZA-5B (**21b**)	138
BMS 185878 (**23a**)	6	21000		NIH3T3/H-RAS and NIH3T3/K-RAS, endogenous RAS more resistant	139, 140, 141

ND: not determined

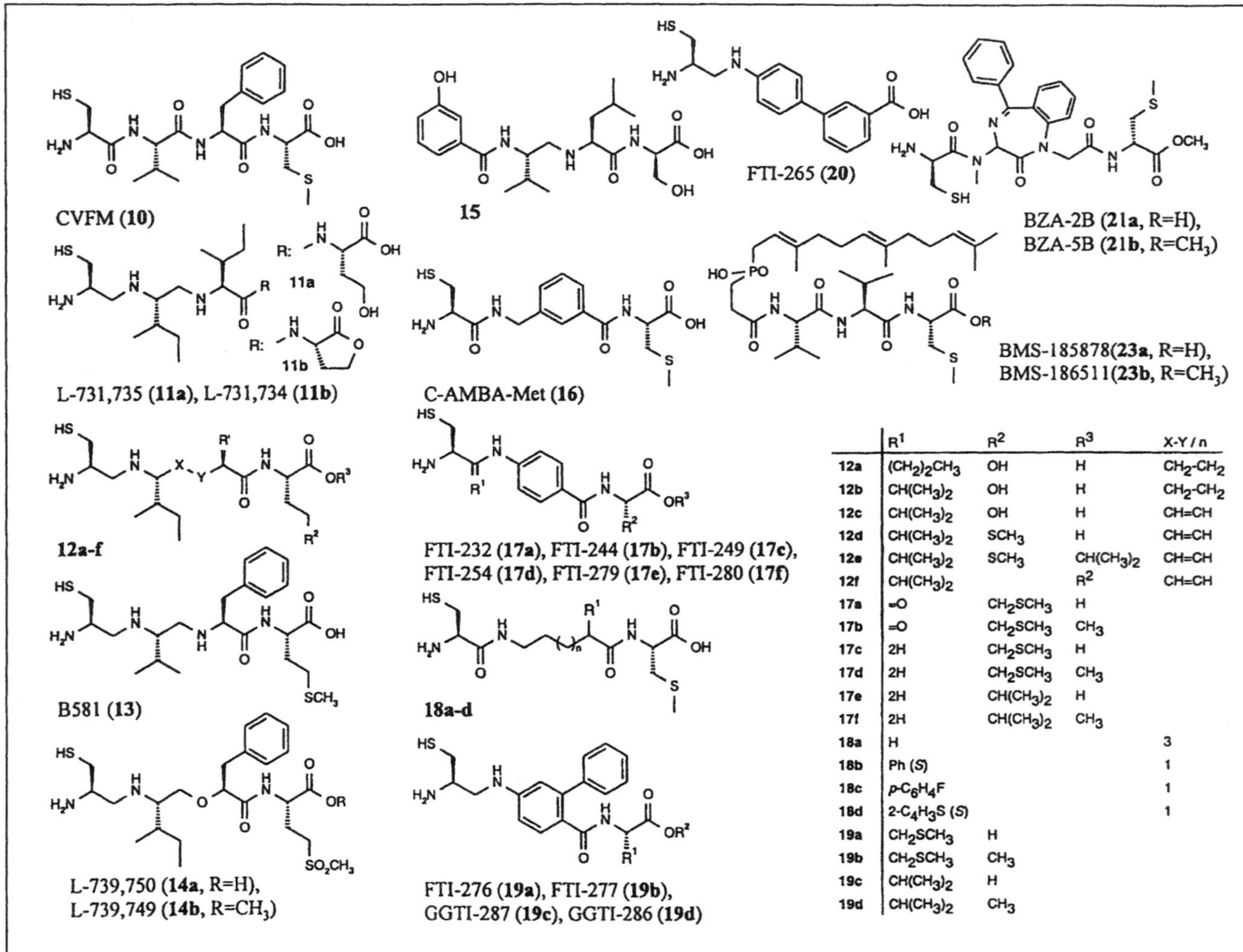

	R^1	R^2	R^3	X-Y / n
12a	$(CH_2)_2CH_3$	OH	H	CH_2-CH_2
12b	$CH(CH_3)_2$	OH	H	CH_2-CH_2
12c	$CH(CH_3)_2$	OH	H	CH=CH
12d	$CH(CH_3)_2$	SCH_3	H	CH=CH
12e	$CH(CH_3)_2$	SCH_3	$CH(CH_3)_2$	CH=CH
12f	$CH(CH_3)_2$		R^2	CH=CH
17a	=O	CH_2SCH_3	H	
17b	=O	CH_2SCH_3	CH_3	
17c	2H	CH_2SCH_3	H	
17d	2H	CH_2SCH_3	CH_3	
17e	2H	$CH(CH_3)_2$	H	
17f	2H	$CH(CH_3)_2$	CH_3	
18a	H			3
18b	Ph (*S*)			1
18c	p-C_6H_4F			1
18d	2-C_4H_3S (*S*)			1
19a	CH_2SCH_3	H		
19b	CH_2SCH_3	CH_3		
19c	$CH(CH_3)_2$	H		
19d	$CH(CH_3)_2$	CH_3		

Fig. 4.6. Structures of peptidomimetic compounds and a bisubstrate analog that inhibit protein prenylation.

12a-f are examples from an extensive study on peptidomimetics in which the central peptide bond was replaced by a two carbon tether.[130] These compounds illustrate how slight structural changes can affect the activity. The 20-fold loss in activity between **12a** and the analog **12b** is triggered by a minor difference in the configuration of amino acid residue A_2 of the CA_1A_2X motif. Oxidation of the carbon tether to a double bond led to inhibitors with better potency (**12c-f**). Recently, an analog of B581, called B956, has been reported in which the linkers between the first two side chains are replaced by olefinic units.[131]

A step further in terms of an increasing degree of peptide modifications is the replacement of whole units of the peptide structure. The replacement of the N-terminal cysteine as in **15** by a phenol moiety which was expected to emulate certain chemical properties of the sulhydryl residue is detrimental to the inhibitory activity.[132] The replacement of the central peptide moiety by an aromatic ring system turned out to be a very successful approach in creating peptide analogs without any peptidic bonds. A number of compounds have been reported with this kind of scaffold which exhibit good to very good potencies.[133,134] The regiochemistry of the central aromatic unit and its conformational impact on the molecule seems to be less crucial than originally assumed, as the regioisomers C-AMBA-Met (**16**) and FTI-232 (**17a**) are of similar inhibitory potency.[135,136] It is remarkable how strongly the activity is enhanced by adding another hydrophobic aromatic ring to the central portion of the molecule; for example FTI-276 (**19a**) has a 600-fold higher inhibitory activity than FTI-249 (**17c**). The switch in inhibitory potency from FTase to GGTase I which is displayed in FTI-276 (**19a**) and GGTI-287 (**19c**) follows the known structural requirements in the C-terminal residue for substrates of the two enzymes.[67,137]

The central portion of the molecule can not only be replaced by a rigid ring system but also by flexible carbon chains (**18a-d**).[133] A series of compounds with unsubstituted linkers of different lengths show a defined dependence of potency and selectivity towards the two prenyltransferases. Addition of various aryl substituents to the chain leads to strong enhancement in activity, once more pointing to the importance of hydrophobic features in this part of the molecule for inhibitory activity. Surprisingly, the stereochemistry of the aryl substitution affected GGTase I activity much more strongly than FTase activity. **18b** and **18d** have close to a 10-fold higher inhibitory activity than their diastereomers with respect to the new stereocenter while the FTase activity remains almost unaffected. It will be interesting to study these kind of features once the three dimensional structure of the protein is known.

A series of benzodiazepine derivatives, of which BZA-2B (**21**) is the most effective, represent another approach to how the central moiety of the peptide can be replaced.[138] The benzodiazepine core was derived from functional analysis because it should provide a favorable spatial arrangement of the cysteine and the C-terminal amino acid. Replacing the N-terminus with other amino acids (Ser, Leu) led to less active inhibitors. An important structural feature of these compounds is the N-methylation which creates a set of diastereomers with a 300-fold difference in inhibitory potency.

The highest degree of backbone replacement in the peptidomimetic inhibitors is displayed by FTI-265 (**20**) which nevertheless has a similar potency as the tetrapeptide templates.[136] It can be applied in vivo without masking of the carboxylate.

Bisubstrate Analogs

A set of compounds were designed by using an active site model in which characteristics of both enzyme substrates were incorporated into the inhibitor molecule.[139,140] The crucial part of these compounds is the central linker unit which mimics the site of the prenylation reaction. The first series of analogs was synthesized with various branched carbon tethers in the linker unit. The concept evolved into another set of compounds in which a phosphinyl- or phosphonyl unit was incorporated into the linker unit. Both types of compounds are equally active and selective. The phosphinyl acid-based prodrug BMS-186511 (**23b**) which is derived from BMS-185878 (**23a**) has been studied for its in vivo activity.[141]

VI. BIOLOGICAL EFFECTS OF PRENYLTRANSFERASE INHIBITORS

Suppression of *RAS*-Activated Phenotypes

The biological activity of some of the above described inhibitors has been studied in different in vivo systems. The systems under study range from mammalian and yeast cell cultures to *RAS* induced tumors in mice; they also include animal models like *C. elegans*, *Xenopus laevis* oocytes and *Drosophila*. In the following, we will discuss the biological activity of FTase inhibitors in general terms. For details on dosages or employed cell systems we refer to Table 4.6 which summarizes results about the in vivo activities of inhibitors.

It has been shown that FTase inhibitors can inhibit cellular RAS farnesylation and processing.[92,117,127,131,136,138,141-143] Typically, RAS processing is probed by immunoblotting with anti-RAS antibodies and detection of gel mobility shift differences between farnesylated and unprocessed protein with [^{35}S]-methionine labeled protein. Unprocessed RAS usually is localized in the cytosol. General protein prenylation commonly appears to be unaffected by FTase inhibitors, as analyzed by looking at the incorporation pattern of [^{3}H]-mevalonate. These results indicate that the specificity for FTase inhibition over GGTase inhibition also applies to in vivo systems. Specifically, an unaffected prenylation status has been demonstrated for Rap1A which gets geranylgeranylated or chimeric RAS forms that have been equipped with the recognition sequence for GGTase I.[117,136,141,143,144]

Furthermore, the inhibition of RAS farnesylation does indeed block the abnormal growth of *RAS* activated cells with effects on the growth rate as well as the altered morphology. In a number of cases, anchorage-independent growth of *RAS* transformed cells in soft-agar was inhibited.[92,117,127,131,138,141,142] It must be noted that differences in in vitro activity of FTase inhibition towards the different forms of RAS are also encountered in vivo. This was clearly indicated in a recent study in which 14 tumor cell lines activated by different forms of *RAS* were treated with the peptidomimetic inhibitor B956.[131] K-*RAS* and N-*RAS* transformed cells required at least a one order of magnitude higher dosage than H-*RAS* activated cells for inhibition of anchorage-independent growth. It is generally accepted that the effects on the transformed phenotype are the result of interference with RAS activity. In several studies *Raf*- or *Mos*-transformed cells which are activated downstream of RAS remain unaffected by FTase inhibitors.[127,138,143,144] On the other hand, *Src* transformations can also be suppressed by FTase inhibitors.[138] The occurrence of growth inhibition in non-*RAS* transformed cell lines indicates additional modes of action for FTase inhibition and will be discussed below. Finally, we want to point out that the required dosage for cellular activity is typically much higher than the one indicated by the in vitro potency.

A closer look on the mechanistic level confirms the assumption that inhibition of RAS farnesylation blocks RAS action. In different cases FTase inhibitors disrupt downstream events of the MAP kinase cascade which eventually results in blocking of the transcriptional activity of Jun and Elk-1.[143,145,146] It is important to note that this interference with the MAP kinase cascade only takes place in cells that are transformed with *RAS*, and only if RAS carries the farnesylation sequence. This finding points to the idea that nonfarnesylated activated RAS acts as dominant inhibitor over farnesylated RAS[147] (see following section). However, in untransformed cells FTI-244 was shown to interfere with PDGF-induced DNA synthesis which again points to additional targets of prenylation inhibition.[146]

There are a number of cell models that show the effects of FTase inhibitors on *RAS* mediated signaling. In yeast, RAS activates the cAMP pathway, and when activated, results in phenotypes such as heat shock sensitivity. This activated phenotype can be suppressed by the addition of manumycin.[56] In *C. elegans*, the expression of activated *RAS* causes a multi-vulva phenotype which can also be suppressed by manumycin.[104] *Drosophila* transgenic lines bearing the activating $RAS1^{val12}$ mutation exhibit roughening of the external surface of the eye. The FTase inhibitor FTI-280, which is also an inhibitor of GGTase I, displays efficient suppression of this rough-eye phenotype.[67] This is expected, since GGTase I is involved in the modification of this particular

Table 4.6. In vivo effects of prenyltransferase inhibitors

Compound	In Vivo Activity	Refs.
Gliotoxin	- suppression of the RAS mediated multivulva phenotype in *C. elegans* - suppression of yeast gpa1	104,105
Manumycin	- suppression of yeast gpa1 - suppression of RAS activation in yeast - suppression of the RAS mediated multivulva phenotype in *C. elegans* - suppression of tumor growth of fibrosarcoma (K-BALB) in mice (63% at 6.3 mg/kg x day)	55,56, 104,105a
SCH44342 (**22**)	- suppression of transformed morphology in Cos/H-RAS (100% at 4.6 μM), but not of cells activated by chimeric RasGG or RAS-SVLS - toxicity: acetate incorporation into sterols affected at 50 mM (50%) - no interference with PDGF activation of MAP kinase in NIH3T3 cells	117
L-731,734 (**11b**)	- suppression of anchorage-independent growth of Rat1/RAS at 1mM and also of Rat1/Src, but not of *Raf-* or *Mos*-transformed cells	126,127,129
B581 (**13**)	- reversion of transformed morphology of NIH3T3/H-RAS and also in case of transformation by chimeric RasGG and myr-RAS; the reversed morphology is distinct from that of untransformed cells. - inhibition of anchorage-independent growth of NIH3T3/H-RAS (no effect on transformed cells with RasGG or myr-RAS) - disruption of induced MAP kinase activation and transcriptional activity of Jun and Elk-1 - blocking of H-RAS induced maturation of microinjected *Xenopus laevis* oocytes in a dose-dependent manner (7h at 25 pmol)	128,143
B956/B1086	- inhibition of anchorage-independent growth of 14 human RAS-activated tumor cell lines (100% at 0.2 - 60 μM depending on the form of RAS and the cell type) - inhibition of anchorage-independent growth of 5 human tumor cell lines without RAS activation (100% at 10 - 80 μM) - in nude mice growth inhibition of implanted-activated human tumors (EJ-1 bladder carcinoma (H-*RAS*) >> H1080 fibrosarcoma (N-*RAS*) > HCT116 colon carcinoma (K-*RAS*) at 100 mg/kgxday); no gross toxicity	131
	- inhibition of tumor growth (NIH3T3/H-RAS) correlates with accumulation of cytosolic unprocessed RAS	131

L-739,749 (**14b**)	- reversion of transformed morphology in Rat1/RAS cells (100% at 10 μM) without correlation to RAS processing; the reversion takes place within 24 hours and is stable for up to 10 days after single addition - reduction of anchorage-dependent monolayer growth rate of Rat1/RAS by continuous treatment (77% reduction at 50 μM; 22% reduction in untransformed cells). - inhibition of anchorage-independent growth of Rat1/H-RAS (100% at 10 μM), Rat1/K-RAS and Rat1/N-RAS - enlargement of transformed and normal Rat1 cells - induction of actin stress fibers in normal and H-RAS transformed Rat1 cells, also in Rat1/Src and T24 human bladder carcinoma cells - inhibition of anchorage-independent growth of human pancreatic adenocarcinoma cell line PSN-1 (100% at >100 μM) - in nude mice growth inhibition of implanted *RAS*-dependent tumors (Rat1/hu-RAS) by 66% (H-RAS), 58% (N-RAS) and 51% (K-RAS), 20 mg/kg x day, 5 days); no effect on *Raf*- or *Mos*-transformed tumors (Rat1) - regression of mammary and salivary tumors in transgenic mice (H-RAS) by L-744,832 (isopropylester-analog of **14b**; 40 mg/kg x day)	129,142, 144,150,153
FTI-244 (**17a**) FTI-254 (**17d**) FTI-280 (**17f**) FTI-277 (**19b**) GGTI-286 (**19d**) FTI-265 (**20**)	- **17a**: in NIH3T3 blocking of PDGF (platelet derived growth factor) induced DNA synthesis at 200 - 400 μM, but no interference of PDGF stimulation of MAP kinase activity. Significantly reduced phosphorylation of PDGFR; interference with PI-3K association to PDGFR. - **19b**: disruption of RAS induced MAP kinase activation in NIH3T3/H-RAS (100% at 1 μM) but not in NIH3T3/Raf or transformants with chimeric RAS-GG or myr-RAS. - **19d**: disruption of *RAS* induced MAP kinase activation in NIH3T3/K-RAS4B (100% at 10 μM); inhibition of rap1A geranylgeranylation; cytotoxic above 30 μM (rounded cell morphology) - **19b**: blocking of activated H-RAS in NIH3T3 results in cytosolic RAS/Raf complexes, in which Raf kinase is inactive - in *Drosophila* larvae suppression of supernumerary formation of R7 photoreceptor cells after ras1 activation (**17f>17d>17a, 20**) - **17d**: no effect on transformed flies with activated sevenless receptor tyrosine kinase, Raf kinase or chimeric myr-RAS, photoreceptor phenotype of untransformed larvae is almost unaffected; no obvious lethal or abnormal developmental effects on *Drosophila* - **17f**: in *Drosophila* suppression of the roughened phenotype of activated rap1A - **19b**: in nude mice blocking of tumor growth of human lung carcinoma (Calu-1; K-*RAS* and *p53* dependent) by 50 mg/kg/day over 36 days). Inhibition of tumor growth (NIH3T3/H-RAS) correlates with accumulation of cytosolic unprocessed RAS. No effect on xenografts with activated Raf non-*RAS* dependent tumor cell line (NCI-810); no gross toxicity	67,135, 136,137, 146,147

Continued...

Table 4.6. In vivo effects of prenyltransferase inhibitors (continued)

Compound	In Vivo Activity	Refs.
BZA-5B (**21**)	- reversion of transformed morphology in rat1/H-RAS cells (100% at 200 μM) and in NIH3T3/K-RasB; no effect on *Src* activation - inhibition of cell growth of rat1/H-RAS (~90% at 25 μM) and of mouse myoblasts C41/H-RAS (75 μM), no effect on *Src*-activated or untransformed cells - disruption of MAPK cascade in rat1/H-RAS at concentrations but not in cells transformed with wildtype H-RAS or activated chimeric RasGG (100 μM), at concentration used for MAPK disruption RAS processing is only partially blocked - in untransformed cells no effect on MAP pathway and no prevention of acute activation by EGF, although no farnesylated H-RAS detectable - despite complete inhibition of protein farnesylation (25 μM, IC_{50} = 7.5 μM) assembly of the nuclear lamina and membrane association of RAS (CHO-K1) - no inhibition of neuronal differentiation as response to NGF in PC-12 cells	19,130, 138,145,156
BMS 186511 (**23b**)	- reversion of transformed morphology in NIH3T3/H-RAS (100% at 100 μM) in 4-5 days - inhibition of anchorage-independent growth of H-RAS (100%) and K-RAS (64%) transformed NIH3T3 cells (at 100 μM) - geranylgeranylation affected after long-term treatment (4 days, 100 μM) - restoration of actin cytoskeleton in NIH3T3/H-RAS (100% at 100 μM) - dose dependent inhibition of malignant growth of schwannoma cells from NF1 patient (ST88-14) - no effect on NIH3T3 transformed with oncogenic R-RAS2/TC21 - blocking of H-RAS induced maturation of microinjected *Xenopus laevis* oocytes	139,140, 141,151,155

RAS (see section III). Another model system for studying RAS mediated signaling utilizes *Xenopus* oocytes in which activated RAS causes maturation (germinal vesicle breakdown).[148] FTase inhibitors were shown to block this *RAS* induced maturation process.[128,141]

Moreover, suppression of H-*RAS* transformed phenotypes can effectively be achieved with FTase inhibitors in *RAS* tumor xenografts in nude mice or more recently in transgenic mice. Mice injected subcutaneously with oncogenic *RAS* were treated with L-739,749 which resulted in a significant inhibition of tumor growth.[144] Similarly, FTI-276 inhibited in nude mice the growth of human lung carcinoma which contains K-*RAS* and a p53 deletion.[149] The extent of tumor growth inhibition correlates with the accumulation of unprocessed RAS. This correlation was also shown with the inhibitor B956 and its prodrug derivative B1086.[131] This study also demonstrated the striking differences in growth inhibition of xenografts containing different forms of *RAS*. The N-*RAS* derived HT-1080 fibrosarcoma was significantly more resistant to B1086 than the H-*RAS* induced EJ-1 bladder carcinoma cell line. The K-*RAS* derived HCT116 colon carcinoma cell line showed the least sensitivity although in vitro results indicated that it had a greater sensitivity to the inhibitor than the HT-1080 cell line. The in vivo potency of FTase inhibitors has also been demonstrated in transgenic mice which overexpress v-H-*RAS*.[150] L-739,749 causes regression of mammary and salivary tumors; however a continuous supply of the inhibitor was required, as removing the inhibitor resulted in the reappearance of the tumor.

Issues Raised by the Studies Utilizing FTase Inhibitors

The above observations suggest that inhibitors of farnesyltransferase inhibit the growth of *RAS*-transformed mammalian cells. Studies with mouse models provide convincing evidence that inhibitors of RAS prenylation hold the potential to be developed as therapeutics against *RAS*-dependent forms of cancer. In addition, FTase inhibitors could also be useful for the treatment of other diseases that are related to malfunctions of cellular signaling pathways involving RAS. One such case is neurofibromatosis type 1 (NF1), a genetic disease of the peripheral nervous system in which a defective or absent NF1 protein is no longer able to act as GAP (GTPase activating protein) which downregulates RAS. FTase inhibitors can block malignant growth of schwannoma cells from an NF1 (neurofibromatosis type 1) patient.[151]

The cellular responses to the inhibition of RAS farnesylation may originate from the accumulation of oncogenic RAS in the cytosol. Treatment of transformed cells with FTI-276 resulted in the accumulation of the RAS-Raf complex in the cytosol resulting in the loss of Raf activity.[147] Thus, nonprenylated cytosolic RAS could sequester Raf and exhibit a dominant negative effect. Because the oncogenic RAS protein stays in a GTP-bound state, the formation of the inactive complex occurs very efficiently. In contrast, normal RAS proteins are quickly converted to the GDP-bound form which does not favor the complex formation.[152] This may also be part of the reason for the absence of significant effects of FTase inhibitors on untransformed cells.

The various studies on FTase inhibitors resulted in a number of unexpected observations that do not fit into the current model of RAS action or other general features of cell signaling. One aspect concerns the effect of FTase inhibitors on the morphology of transformed cells. Morphological reversion of transformed cells has been observed by the administration of FTase inhibitors. However, this effect does not necessarily correlate with inhibition of RAS farnesylation.[142] Furthermore, morphological reversion, in one study, was also observed in cells that were either transformed with the *Raf* oncogene or chimeric *RAS* that gets modified by geranylgeranylation or myristoylation.[143] In a similar manner, inhibition of soft-agar growth of tumor cell lines that bear non-*RAS* transformations was observed following FTase inhibitor treatment.[131] The required dosage for this unexpected effect was of the same order of magnitude as that required for K-*RAS* transformed cell lines. These findings suggest that the biological effect of FTase inhibition may involve pathways in addition to those depending on RAS. Differences in sensitivity exhibited by different cell lines expressing the same *RAS* oncogene (K-*RAS*) also suggests multiple responses to FTase inhibition that might vary between different cell types. A likely candidate which could be responsible for the additional effects of inhibition is RhoB, which is known to undergo farnesylation as well as geranylgeranylation (see section III).[72] Careful observation shows that the morphological reversion of transformed cells involves cell enlargement and actin stress fiber induction.[142] This strongly suggests that the inhibition of farnesylation affects regulation of the actin cytoskeleton which makes RhoB a possible target.[142,153]

Considering the biological effects discussed above, it is particularly striking that untransformed cells appear to be generally unaffected by FTase inhibitors. Because RAS proteins are essential for growth, one might expect that FTase inhibition would result in a general growth inhibition. But no significant toxic effects on the growth of untransformed cells have been reported and also in mouse models remarkably few toxic effects are seen with peptidomimetic inhibitors. Several reasons could account for these observations. One possibility is that other prenyltransferases may modify RAS in the absence of FTase activity. In fact, yeast cells with the deletion of FTase genes can still grow, and this appears to be due to the modification of RAS by GGTase I (see section III). Another possibility is that in the absence of functional RAS, other G-proteins can activate the MAP kinase pathway. If this is the case, R-RAS should be considered as a possible candidate since its modification is not affected by FTase inhibitors.[154,155] An alternative mechanism for RAS processing might also explain the unexpected observation that BZA-5B does not affect RAS association with the membrane in CHO-K1 cells despite complete inhibition of its farnesylation.[156]

Yet another possible explanation for the lack of effects on untransformed cells has to do with the fact that the prenylation of K-RAS4B is not affected by FTase inhibitors.[19] As discussed before, this form of K-RAS that contains a polybasic domain near the C-terminus can be modified by GGTase I.[19] Thus, its modification might resist the presence of FTase inhibitors, thereby enabling cell growth. Because K-*RAS* is the most prevalent type of *RAS* oncogene, it is crucial to address this question in the hopes of developing a therapeutic concept. As a "byproduct" of the development of FTase inhibitors, a number of compounds have been identified that are effective in the inhibition of GGTase I. These compounds may serve as starting points for further exploration in this direction.[67,126,133,136,138] In a recent study, GGTI-286, an inhibitor of GGTase I was shown to inhibit the K-*RAS*4B activation of MAP kinase.[137] The required concentration of 10 μM was not toxic to cells; however some signs of toxicity were seen at a three-fold higher concentration. Toxicity may turn out to be a bigger problem with inhibitors of GGTase I, as opposed to those of FTase, since geranylgeranylated proteins are more abundant in the cell than farnesylated ones. Further characterization will be needed to evaluate the potential of GGTase I inhibitors.

From the above it is obvious that inhibitors of prenyltransferases are not only of interest for their potential as future anticancer drugs but that they are also valuable tools for studying cellular physiology and signaling. Inhibitors of prenylation may also find applications for cellular events other than the *RAS* network that involve prenyltransferases. For example, recently it was found that the α-subunit of FTase associates with the cytoplasmic domain of the TGF-β I (transforming growth factor-β) type 1 receptor.[157,158] This raises the possibility that FTase or GGTase I plays a role in the regulation of TGF-β mediated signaling events.

Acknowledgments

We wish to thank Drs. Hiroshi Mitsuzawa, Mitsunobu Hara and Hirofumi Nakano for helpful discussions. Nicole Robinson provided valuable suggestions about the manuscript. We wish to thank Prof. Dr. Axel Zeeck, in particular, for his continued interest and helpful discussions. This work is supported by NIH grant CA41996.

References

1. Lowy DR, Willumsen BM. Function and regulation of RAS. Annu Rev Biochem 1993; 62:851-91.
2. Tamanoi F. Inhibitors of RAS farnesyltransferases. Trends Biochem Sci 1993; 18:349-53.
3. Gibbs JB, Oliff A, Kohl NE. Farnesyltransferase inhibitors: RAS research

yields a potential cancer therapeutic. Cell 1994; 77:175-8.

4. Zhang FL, Casey PJ. Protein prenylation: molecular mechanisms and functional consequences. Annu Rev Biochem 1996; 65:241-270.
5. Buss JE, Marsters JC Jr. Farnesyltransferase inhibitors: the successes and surprises of a new class of potential cancer chemotherapeutics. Chem Biol 1995; 2:787-91.
6. Caldwell GA, Naider F, Becker JM. Fungal lipopeptide mating pheromones: a model system for the study of protein prenylation. Microbiol Rev 1995; 59:406-22.
7. Omer CA, Gibbs JB. Protein prenylation in eukaryotic microorganisms: genetics, biology and biochemistry. Mol Microbiol 1994; 11:219-25.
8. Glomset JA, Farnsworth CC. Role of protein modification reactions in programming interactions between RAS-related GTPases and cell membranes. Annu Rev Cell Biol 1994; 10:181-205.
9. Schafer WR, Rine J. Protein prenylation: genes, enzymes, targets, and functions. Annu Rev Genet 1992; 26:209-37.
10. Clarke S. Protein isoprenylation and methylation at carboxyl-terminal cysteine residues. Annu Rev Biochem 1992; 61:355-86.
11. Fujiyama A, Tamanoi F. Processing and fatty acid acylation of RAS1 and RAS2 proteins in *Saccharomyces cerevisiae.* Proc Natl Acad Sci USA 1986; 83:1266-70.
12. Farh L, Mitchell DA, Deschenes RJ. Farnesylation and proteolysis are sequential, but distinct steps in the CaaX box modification pathway. Arch Biochem Biophys 1995; 318:113-21.
13. Fujiyama A, Tsunasawa S, Tamanoi F et al. S-farnesylation and methyl esterification of C-terminal domain of yeast RAS2 protein prior to fatty acid acylation. J Biol Chem 1991; 266:17926-31.
14. Casey PJ, Solski PA, Der CJ et al. P21RAS is modified by a farnesyl isoprenoid. Proc Natl Acad Sci USA 1989; 86:8323-7.
15. Hancock JF, Magee AI, Childs JE et al. All RAS proteins are polyisoprenylated but only some are palmitoylated. Cell 1989; 57:1167-77.
16. Shabinian S, Silvius JR. Doubly-lipid-modified protein sequence motifs exhibit long-lived anchorage to lipid bilayer membrane. Biochemistry 1995; 34:3813-22.
17. Hancock JF, Paterson H, Marshall CJ. A polybasic domain or palmitoylation is required in addition to the CAAX motif to localize p21RAS to the plasma membrane. Cell 1990; 63:133-9.
18. Ghomashchi F, Zhang X, Liu L et al. Binding of prenylated and polybasic peptides to membranes: affinities and intervesicle exchange. Biochemistry 1995; 34:11910-8.
19. James GL, Goldstein JL, Brown MS. Polylysine and CVIM sequences of K-RASB dictate specificity of prenylation and confer resistance to benzodiazepine peptidomimetic in vitro. J Biol Chem 1995; 270:6221-6.
20. Willumsen BM, Christensen A, Hubbert NL et al. The p21 RAS C-terminus is required for transformation and membrane association. Nature 1984; 310:583-6.

21. Deschenes RJ, Broach JR. Fatty acylation is important but not essential for *Saccharomyces cerevisiae* RAS function. Mol Cell Biol 1987; 7:2344-51.
22. Fujiyama A, Matsumoto K, Tamanoi F. A novel yeast mutant defective in the processing of RAS proteins: assessment of the effect of the mutation on processing steps. EMBO J 1987; 6:223-8.
23. Powers S, Michaelis S, Broek D et al. RAM, a gene of yeast required for a functional modification of RAS proteins and for production of mating pheromone a-factor. Cell 1986; 47:413-22.
24. Itoh T, Kaibuchi K, Masuda T et al. The post-translational processing of RAS p21 is critical for its stimulation of mitogen-activated protein kinase. J Biol Chem 1993; 268:3025-8.
25. McGeady P, Kuroda S, Shimizu K et al. The farnesyl group of H-RAS facilitates the activation of a soluble upstream activator of mitogen-activated protein kinase. J Biol Chem 1995; 270:26347-51.
26. Kuroda Y, Suzuki N, Kataoka T. The effect of posttranslational modifications on the interaction of RAS2 with adenylyl cyclase. Science 1993; 259: 683-6.
27. Reiss Y, Goldstein JL, Seabra MC et al. Inhibition of purified p21*RAS* farnesyl:protein transferase by Cys-AAX tetrapeptides. Cell 1990; 62:81-8.
28. Moores SL, Schaber MD, Mosser SD et al. Sequence dependence of protein isoprenylation. J Biol Chem 1991; 266:14603-10.
29. Reiss Y, Stradley SJ, Gierasch LM et al. Sequence requirement for peptide recognition by rat brain $p21^{ras}$ protein farnesyltransferase. Proc Natl Acad Sci USA 1991; 88:732-6.
30. Gomez R, Goodman LE, Tripathy SK et al. Purified yeast protein farnesyltransferase is structurally and functionally similar to its mammalian counterpart. Biochem J 1993; 289:25-31.
31. Mayer MP, Prestwich GD, Dolence JM et al. Protein farnesyltransferase: production in *Escherichia coli* and immunoaffinity purification of the heterodimer from *Saccharomyces cerevisiae.* Gene 1993; 132:41-7.
32. Moomaw JF, Casey PJ. Mammalian protein geranylgeranyltransferase. Subunit composition and metal requirements. J Biol Chem 1992; 267:17438-43.
33. Yokoyama K, Gelb MH. Purification of a mammalian protein geranylgeranyltransferase. Formation and catalytic properties of an enzyme-geranylgeranyl pyrophosphate complex. J Biol Chem 1993; 268:4055-60.
34. Finegold AA, Johnson DI, Farnsworth CC et al. Protein geranylgeranyltransferase of *Saccharomyces cerevisiae* is specific for Cys-Xaa-Xaa-Leu motif proteins and requires the CDC43 gene product but not the DPR1 gene product. Proc Natl Acad Sci USA 1991; 88:4448-52.
35. Mayer ML, Caplin BE, Marshall MS. CDC43 and RAM2 encode the polypeptide subunits of a yeast type I protein geranylgeranyltransferase. J Biol Chem 1992; 267:20589-93.
36. Seabra MC, Goldstein JL, Sudhof TC et al. Rab geranylgeranyl transferase. A multisubunit enzyme that prenylates GTP-binding proteins terminating in Cys-X-Cys or Cys-Cys. J Biol Chem 1992; 267:14497-503.

37. Horiuchi H, Kawata M, Katayama M et al. A novel prenyltransferase for a small GTP-binding protein having a C-terminal Cys-Ala-Cys structure. J Biol Chem 1991; 266:16981-4.
38. Kinsella BT, Maltese WA. Rab GTP-binding proteins implicated in vesicular transport are isoprenylated in vitro at cysteines within a novel carboxyl-terminal motif. J Biol Chem 1991; 266:8540-4.
39. Khosravi-Far R, Clark GJ, Abe K et al. RAS (CXXX) and Rab (CC/CXC) prenylation signal sequences are unique and functionally distinct. J Biol Chem 1992; 267:24363-8.
40. Jiang Y, Rossi G, Ferro-Novick S. Bet2p and Mad2p are components of a prenyltransferase that adds geranylgeranyl onto Ypt1p and Sec4p. Nature 1993; 366:84-6.
41. Farnsworth CC, Seabra MC, Ericsson LH et al. Rab geranylgeranyl transferase catalyzes the geranylgeranylation of adjacent cysteines in the small GTPases Rab1A, Rab3A, and Rab5A. Proc Natl Acad Sci USA 1994; 91:11963-7.
42. Andres DA, Seabra MC, Brown MS et al. cDNA cloning of component A of Rab geranylgeranyl transferase and demonstration of its role as a Rab escort protein. Cell 1993; 73:1091-9.
43. Alexandrov K, Horiuchi H, Steele-Mortimer O et al. Rab escort protein-1 is a multifunctional protein that accompanies newly prenylated Rab proteins to their target membranes. EMBO J 1994; 13:5262-73.
44. Chen WJ, Andres DA, Goldstein JL et al. cDNA cloning and expression of the peptide-binding β-subunit of rat p21ras farnesyltransferase, the counterpart of yeast DPR1/RAM1. Cell 1991; 66:327-34.
45. Kohl NE, Diehl RE, Schaber MD et al. Structural homology among mammalian and *Saccharomyces cerevisiae* isoprenyl-protein transferases. J Biol Chem 1991; 266:18884-8.
46. Zhang FL, Diehl RE, Kohl NE et al. cDNA cloning and expression of rat and human protein geranylgeranyltransferase type-I. J Biol Chem 1994; 269:3175-80.
47. Armstrong SA, Seabra MC, Sudhof TC et al. cDNA cloning and expression of the α and β subunits of rat Rab geranylgeranyl transferase. J Biol Chem 1993; 268:12221-9.
48. Seabra MC, Brown MS, Slaughter CA et al. Purification of component A of Rab geranylgeranyl transferase: possible identity with the choroideremia gene product. Cell 1992; 70:1049-57.
49. Cremers FP, Armstrong SA, Seabra MC et al. REP-2, a Rab escort protein encoded by the choroideremia-like gene. J Biol Chem 1994; 269:2111-7.
50. Seabra MC, Ho YK, Anant JS. Deficient geranylgeranylation of Ram/Rab27 in choroideremia. J Biol Chem 1995; 270:24420-7.
51. Goodman LE, Perou CM, Fujiyama A et al. Structure and expression of yeast DPR1, a gene essential for the processing and intracellular localization of RAS proteins. Yeast 1988; 4:271-81.
52. He B, Chen P, Chen SY et al. RAM2, an essential gene of yeast, and

RAM1 encode the two polypeptide components of the farnesyltransferase that prenylates a-factor and RAS proteins. Proc Natl Acad Sci USA 1991; 88:11373-7.
53. Finegold AA, Schafer WR, Rine J et al. Common modifications of trimeric G proteins and RAS protein: involvement of polyisoprenylation. Science 1990; 249:165-9.
54. Nakayama N, Kaziro Y, Arai K et al. Role of STE genes in the mating factor signaling pathway mediated by GPA1 in *Saccharomyces cerevisiae.* Mol Cell Biol 1988; 8:3777-83.
55. Hara M, Akasaka K, Akinaga S et al. Identification of RAS farnesyltransferase inhibitors by microbial screening. Proc Natl Acad Sci USA 1993; 90:2281-5.
56. Mitsuzawa H, Tamanoi F. In vivo assays for farnesyltransferase inhibitors with *Saccharomyces cerevisiae.* Methods Enzymol 1995; 250:43-51.
57. Ohya Y, Goebl M, Goodman LE et al. Yeast CAL1 is a structural and functional homologue to the DPR1 (RAM) gene involved in RAS processing. J Biol Chem 1991; 266:12356-60.
58. Adams AE, Johnson DI, Longnecker RM et al. CDC42 and CDC43, two additional genes involved in budding and the establishment of cell polarity in the yeast *Saccharomyces cerevisiae.* J Cell Biol 1990; 111:131-42.
59. Li R, Havel C, Watson JA et al. The mitotic feedback control gene MAD2 encodes the α-subunit of a prenyltransferase. Nature 1994; 371:438.
60. Fujimura K, Tanaka K, Nakano A et al. The *Saccharomyces cerevisiae* MSI4 gene encodes the yeast counterpart of component A of Rab geranylgeranyltransferase. J Biol Chem 1994; 269:9205-12.
61. Waldherr M, Ragnini A, Schweyer RJ et al. MRS6-yeast homologue of the choroideremia gene. Nature Genet 1993; 3:193-4.
62. Diaz M, Sanchez Y, Bennett T et al. The *Schizosaccharomyces pombe* cwg2+ gene codes for the β-subunit of a geranylgeranyltransferase type I required for β-glucan synthesis. EMBO J 1993; 12:5245-54.
63. Godfrey R, Davey J. Sequence of ptb1, a gene for the β-subunit of the type-II geranylgeranyltransferase from the fission yeast *Schizosaccharomyces pombe.* Yeast, In press.
64. Therrien M, Chang HC, Solomon NM et al. KSR, a novel protein kinase required for RAS signal transduction. Cell 1995; 83:879-88.
65. Yang Z, Cramer CL, Watson JC. Protein farnesyltransferase in plants. Molecular cloning and expression of a homolog of the β-subunit from the garden pea. Plant Physiol 1993; 101 667-74.
66. Morris GN, Pullarkat RK. Bovine brain gray and white matter exhibit differential protein prenyl transferase activity. Neurochem Res 1995; 20:457-60.
67. Kauffmann RC, Qian Y, Vogt A et al. Activated *Drosophila* RAS1 is selectively suppressed by isoprenyl transferase inhibitors. Proc Natl Acad Sci USA 1995; 92:10919-23.
68. Trueblood CE, Ohya Y, Rine J. Genetic evidence for in vivo cross-specificity of the CaaX-box protein prenyltransferases farnesyltransferase and

geranylgeranyltransferase-I in *Saccharomyces cerevisiae.* Mol Cell Biol 1993; 13:4260-75.
69. Ohya, Y, Qadota H, Anraku Y et al. Suppression of yeast geranylgeranyl transferase I defect by alternative prenylation of two target GTPases, Rho1p and Cdc42p. Mol Biol Cell 1993; 4:1017-25.
70. Mitsuzawa H, Esson K, Tamanoi F. Mutant farnesyltransferase β-subunit of *Saccharomyces cerevisiae* that can substitute for geranylgeranyltransferase type I β-subunit. Proc Natl Acad Sci USA 1995; 92:1704-8.
71. Yokoyama K, Goodwin GW, Ghomashchi F et al. A protein geranylgeranyltransferase from bovine brain: implications for protein prenylation specificity. Proc Natl Acad Sci USA 1991; 88:5302-6.
72. Armstrong SA, Hannah VC, Goldstein JL et al. CAAX geranylgeranyl transferase transfers farnesyl as efficiently as geranylgeranyl to RhoB. J Biol Chem 1995; 270:7864-8
73. Adamson P, Marshall CJ, Hall A et al. Post-translational modifications of p21rho proteins. J Biol Chem 1992; 267:20033-8.
74. Goldstein JL, Brown MS, Stradley SJ et al. Nonfarnesylated tetrapeptide inhibitors of protein farnesyltransferase. J Biol Chem 1991; 266:15575-8.
75. Reiss Y, Seabra MC, Armstrong SA et al. Nonidentical subunits of p21H-RAS farnesyltransferase. Peptide binding and farnesyl pyrophosphate carrier functions. J Biol Chem 1991; 266:10672-7.
76. Omer CA, Kral AM, Diehl RE et al. Characterization of recombinant human farnesyl-protein transferase: Cloning, expression, farnesyl diphosphate binding, and functional homology with yeast prenyl-protein transferases. Biochemistry 1993; 32:5167-76.
77. Bukhtiyarov YE, Omer CA, Allen CM. Photoreactive analogs of prenyl diphosphates as inhibitors and probes of human protein farnesyltransferase and geranylgeranyltransferase type I. J Biol Chem 1995; 270:19035-40.
78. Ying W, Sepp-Lorenzino L, Cai K et al. Photoaffinity-labeling peptide substrates for farnesyl-protein transferase and the intersubunit location of the active site. J Biol Chem 1994; 269:470-7.
79. Stradley SJ, Rizo J, Gierasch LM. Conformation of a heptapeptide substrate bound to protein farnesyltransferase. Biochemistry 1993; 32: 12586-90.
80. Koblan KS, Culberson JC, Desolms SJ et al. NMR studies of novel inhibitors bound to farnesyl-protein transferase. Protein Sci 1995; 4:681-8.
81. Reiss Y, Brown MS, Goldstein JL. Divalent cation and prenyl pyrophosphate specificities of the protein farnesyltransferase from rat brain, a zinc metalloenzyme. J Biol Chem 1992; 267:6403-8.
82. Chen WJ, Moomaw JF, Overton L et al. High level expression of mammalian protein farnesyltransferase in a baculovirus system. The purified protein contains zinc. J Biol Chem 1993; 268:9675-80.
83. Vogt A, Sun J, Qian Y et al. Burkitt lymphoma daudi cells contain two distinct farnesyltransferases with different divalent cation requirements. Biochemistry 1995; 34:12398-403.
84. Zhang FL, Moomaw JF, Casey PJ. Properties and kinetic mechanism of

recombinant mammalian geranylgeranyltransferase type I. J Biol Chem 1994; 269:23465-70.
85. Yokoyama K, McGeady P, Gelb MH. Mammalian protein geranylgeranyl-transferase-I: substrate specificity, kinetic mechanism, metal requirements, and affinity labeling. Biochemistry 1995; 34:1344-54.
86. Pompliano DL, Rands E, Schaber MD et al. Steady-state kinetic mechanism of RAS farnesyl:protein transferase. Biochemistry 1992; 31:3800-7.
87. Pompliano DL, Schaber MD, Mosser SD et al. Isoprenoid diphosphate utilization by recombinant human farnesyl:protein transferase: Interactive binding between substrates and a preferred kinetic pathway. Biochemistry 1993; 32:8341-7.
88. Furfine ES, Leban JJ, Landavazo A et al. Protein farnesyltransferase: kinetics of farnesyl pyrophosphate binding and product release. Biochemistry 1995; 34:6857-62.
89. Dolence JM, Poulter CD. A mechanism for posttranslational modifications of proteins by yeast protein farnesyltransferase. Proc Natl Acad Sci USA 1995; 92:5008-11.
89a. Mitsuzawa H, Del Villar K, Tamanoi F. unpublished results, manuscript in preparation.
90. Andres DA, Goldstein JL, Ho YK et al. Mutational analysis of α-subunit of protein farnesyltransferase. Evidence for a catalytic role. J Biol Chem 1993; 268:1383-90.
91. Bolton GL, Sebolt-Leopold JS, Hodges JC. RAS oncogene directed approaches in cancer chemotherapy. Ann Rep Med Chem 1994; 29:165-74.
92. Manne V, Ricca CS, Brown JG et al. RAS farnesylation as a target for novel antitumor agents: potent and selective farnesyl diphosphate analog inhibitors of farnesyltransferase. Drug Dev Res 1995; 34:121-37.
93. Roskoski R Jr, Ritchie P, Gahn LG. Farnesyl-protein transferase and geranylgeranyl-protein transferase assays using phosphocellulose paper absorption. Anal Biochem 1994; 222:275-80.
94. Harwood HJ Jr. Protein farnesyltransferase: measurement of enzymatic activity in 96-well format using TopCount microplate scintillation counting technology. Anal Biochem 1995; 226:268-78.
95. Khan SG, Mukhtar H, Agarwal R. A rapid and convenient filter-binding assay for RAS p21 processing enzyme farnesyltransferase. J Biochem Biophys Methods 1995; 30:133-44.
96. Pompliano DL, Gomez RP, Anthony NJ. Intramolecular fluorescence enhancement: a continuous assay of RAS farnesyl:protein transferase. J Am Chem Soc 1992; 114:7945-6.
97. Cassidy PB, Dolence JM, Poulter CD. Continuous fluorescence assay for protein prenyltransferases. Methods Enzymol 1995; 250:30-43.
98. Omura S, Van Der Pyl D, Inokoshi J et al. Pepticinnamins, new farnesyl -protein transferase inhibitors produced by an actinomycete. I. Producing strain, fermentation, isolation, and biological activity. J Antibiot 1993; 46:222-8.
99. Lingham RB, Silverman KC, Bills GF et al. *Chaetomella acutiseta* pro-

duces chaetomellic acids A and B which are reversible inhibitors of farnesyl-protein transferase. Appl Microbiol Biotechnol 1993; 40:370-4.

100. Gibbs JB, Pompliano DL, Mosser SD et al. Selective inhibition of farnesyl-protein transferase blocks RAS processing in vivo. J Biol Chem 1993; 268:7617-20.

101. Singh SB, Zink DL, Lisch JM et al. Isolation and structure of chaetomellic acids A and B from *Chaetomella acutiseta*: farnesyl pyrophosphate mimic inhibitors of RAS farnesyl-protein transferase. Tetrahedron 1993; 49:5917-26.

102. Dufresne C, Wilson KE, Singh SB et al. Zaragozic acids D and D2: potent inhibitors of squalene synthase and of RAS farnesyl-protein transferase. J Nat Prod 1993; 56:1923-9.

103. Liu WC, Barbacid M, Bulgar M et al. 10'-Desmethoxystreptonigrin, a novel analog of streptonigrin. J Antibiot 1992; 45:454-7.

104. Hara M, Han M. RAS farnesyltransferase inhibitors suppress the phenotype resulting from an activated *RAS* mutation in *Caenorhabditis elegans*. Proc Natl Acad Sci USA 1993; 92:3333-7.

105. Van der Pyl D, Inokoshi J, Shiomi K et al. Inhibition of farnesyl-protein transferase by gliotoxin and acetylgliotoxin. J Antibiot 1992; 45:1802-5.

105a. Nagase T, Kawata S, Tamura S et al. Inhibition of Cell Growth of Human Hepatoma Cell Line (Hep G2) By a Farnesyl Protein Transferase Inhibitor: A Preferential Suppression of fas Farnesylation J Cancer 1996; 65:620-626.

106. Sattler I, Gröne C, Zeeck A. New compounds of the manumycin group of antibiotics and a facilitated route for their structure elucidation. J Org Chem 1993; 58:6583-7.

107. Zeeck A, Schröder K, Frobel K et al. The structure of manumycin. J Antibiot 1987; 40:1530-40.

108. Singh SB, Liesch JM, Lingham RB et al. Actinoplanic Acid A: A macrocyclic polycarboxylic acid which is a potent inhibitor of RAS farnesyl-protein transferase. J Am Chem Soc 1994; 116:11606-7.

109. Jayasuriya H, Ball RG, Zink DL et al. Barceloneic acid A, a new farnesyl -protein transferase inhibitor from a Phoma species. J Nat Prod 1995; 58:986-91.

110. Kaneko T, Dabrah T, Harwood H et al. Novel inhibitors of squalene synthase and protein farnesyltransferase. posterpresentation: "Conference on biotechnology of microbial products". Oiso, Japan, April 1995.

111. Singh SB, Zink DL, Bills GF et al. Cylindrol A: a novel inhibitor of RAS farnesyl-protein transferase from *Cylindrocarpon lucidum*. Tetrahedron Lett 1995; 36:4935-8.

112. Singh SB, Jones ET, Goetz MA et al. Fusidienol: a novel inhibitor of RAS farnesyl-protein transferase from *Fusidium griseum*. Tetrahedron Lett 1994; 35:4693-6.

113. Singh SB, Zink DL, Liesch JM et al. Preussomerins and deoxypreussomerins: novel inhibitors of RAS farnesyl-protein transferase. J Org Chem 1994; 59:6296-302.

114. Van der Pyl D, Cans P, Debernard JJ et al. RPR113228, a novel farnesyl-protein transferase inhibitor produced by *Chrysosporium lobatum*. J Antibiot 1995; 48:736-7.
115. Schulz S, Nyce JW. Inhibition of protein farnesyltransferase: a possible mechanism of tumor prevention by dehydroepiandrosterone sulfate. Carcinogenesis 1994; 15:2649-52.
116. Phife DW, Patton RW, Berrie R et al. SCH 58450, a novel farnesyl protein transferase inhibitor possessing a 6a,12a:7,12-diepoxybenz[a]anthracene ring system. Tetrahedron Lett 1995; 36:6995-8.
117. Bishop WR, Bond R, Petrin J et al. Novel tricyclic Inhibitors of farnesyl protein transferase. J Biol Chem 1995; 270:30611-8.
118. Crowell PL, Lin S, Vedejs E et al. Identification of metabolites of the antitumor agent d-limonene capable of inhibiting protein isoprenylation and cell growth. Canc Chemother Pharmacol 2992; 31:205-12.
119. Gelb MH, Tamanoi F, Yokoyama K et al. The inhibition of protein prenyltransferases by oxygenated metabolites of limonene and perillyl alcohol. Cancer Lett 1995; 91:169-75.
120. Patel DV, Schmidt RJ, Biller SA et al. Farnesyl diphosphate-based inhibitors of RAS farnesyl protein transferase. J Med Chem 1995; 38:2906-21.
121. Kothapalli R, Guthrie N, Chambers AF et al. Farnesylamine: an inhibitor of farnesylation and growth of *RAS*-transformed cells. Lipids 1993; 28:469-73.
122. Cohen LH, Valentijn ARPM, Roodenburg L et al. Different analogs of farnesyl pyrophosphate inhibit squalene synthase and protein:farnesyltransferase to different extents. Biochem Pharmacol 1995; 49:839-45.
123. Valentijn ARPM, van den Berg O, van der Marel GA et al. Synthesis of pyrophosphonic acid analogs of farnesyl pyrophosphate. Tetrahedron 1995; 51:2099-108.
124. Brown MS, Goldstein JL, Paris KJ et al. Tetrapeptide inhibitors of protein farnesyltransferase: Amino-terminal substitution in phenylalanine-containing tetrapeptides restores farnesylation. Proc Natl Acad Sci USA 1992; 89:8313-6.
125. Leftheris K, Kline T, Natarajan S et al. Peptide based p21ras farnesyl transferase inhibitors: systematic modification of the tetrapeptide CA1A2X motif. Bioorg Med Chem Lett 1994; 4:887-92.
126. Graham SL, deSolms SJ, Giuliani EA et al. Pseudopeptide inhibitors of RAS farnesyl-protein transferase. J Med Chem 1994; 37:725-32.
127. Kohl NE, Mosser SD, deSolms SJ et al. Selective inhibition of *RAS*-dependent transformation by a farnesyltransferase inhibitor. Science 1993; 260:1934-7.
128. Garcia AM, Rowell C, Ackermann K et al. Peptidomimetic inhibitors of RAS farnesylation and function in whole cells. J Biol Chem 1993; 268:18415-8.
129. Kohl NE, Conner MW, Gibbs JB et al. Development of inhibitors of protein farnesylation as potential chemotherapeutic agents. J Cell Biochem 1995; Suppl 22:145-50.

130. Wai JS, Bamberger DL, Fisher TE et al. Synthesis and biological activity of RAS farnesyl protein transferase inhibitors tetrapeptide analogs with amino methyl and carbon linkages. Bioorg Med Chem 1994; 2:939-47.
131. Nagasu T, Yoshimatsu K, Rowell C et al. Inhibition of human tumor xenograft growth by treatment with the farnesyl transferase inhibitor B956. Cancer Res 1995; 55:5310-4.
132. Patel DV, Patel MM, Robinson SS et al. Phenol based tripeptide inhibitors of RAS farnesyl protein transferase. Bioorg Med Chem Lett 1994; 4:1883-8.
133. Harrington EM, Kowalczyk JJ, Pinnow SL et al. Cysteine and methionine linked by carbon pseudopeptides inhibit farnesyl transferase. Bioorg Med Chem Lett 1994; 4:2775-80.
134. Nigam M, Seong CM, Qian Y et al. Potent inhibition of human tumor p21ras farnesyltransferase by A1A2-lacking p21ras CA1A2X peptidomimetics. J Biol Chem 1993; 268:20695-8.
135. Qian Y, Blaskovich MA, Saleem M et al. Design and structural requirements of potent peptidomimetic inhibitors of p21ras farnesyltransferase. J Biol Chem 1994; 269:12410-3.
136. Vogt A, Qian Y, Blaskovich MA et al. A non-peptide mimetic of RAS-CAAX: selective inhibition of farnesyltransferase and RAS processing. J Biol Chem 1995; 270:660-4.
137. Lerner EC, Qian Y, Hamilton AD et al. Disruption of oncogenic K-RAS4B processing and signaling by a potent geranylgeranyltransferase I inhibitor. J Biol Chem 1995; 270:26770-3.
138. James GL, Goldstein JL, Brown MS et al. Benzodiazepine peptidomimetics: potent inhibitors of RAS farnesylation in animal cells. Science 1993; 60:1937-42.
139. Bhide, RS, Patel DV, Patel MM et al. Rational design of potent carboxylic acid based bisubstrate inhibitors of RAS farnesyl protein transferase. Bioorg Med Chem Lett 1994; 4:2107-12.
140. Patel DV, Gordon EM, Schmidt RJ et al. Phosphinyl acid-based bisubstrate analog inhibitors of farnesyl protein transferase. J Med Chem 1995; 38:435-42.
141. Manne V, Yan N, Carboni JM et al. Bisubstrate inhibitors of farnesyltransferase: a novel class of specific inhibitors of *RAS* transformed cells. Oncogene 1995; 10:1763-79.
142. Prendergast GC, Davide JP, deSolms SJ et al. Farnesyltransferase inhibition causes morphological reversion of *RAS*-transformed cells by a complex mechanism that involves regulation of the actin cytoskeleton. Mol Cell Biol 1994; 14:4193-202.
143. Cox AD, Garcia AM, Westwick JK et al. The CAAX peptidomimetic compound B581 specifically blocks farnesylated, but not geranylgeranylated or myristylated, oncogenic RAS signaling and transformation. J Biol Chem 1994; 269:19203-6.
144. Kohl NE, Wilson FR, Mosser SD et al. Protein farnesyltransferase inhibitors block the growth of *RAS*-dependent tumors in nude mice. Proc Natl

Acad Sci USA 1994; 91:9141-5.
145. James GL, Brown MS, Cobb MH et al. Benzodiazepine peptidomimetic BZA-5B interrupts the MAP kinase activation pathway in H-RAS-transformed rat-1 cells, but not in untransformed cells. J Biol Chem 1994; 269:27705-14.
146. McGuire TF, Qian Y, Blaskovich MA et al. CAAX peptidomimetic FTI-244 decreases platelet-derived growth factor receptor tyrosin phosphorylation levels and inhibits stimulation of phosphatidylinositol 3-kinase but not mitogen-activated protein kinase. Biochem Biophys Res Comm 1995; 214:295-303.
147. Lerner EC, Qian Y, Blaskovich MA et al. RAS CAAX peptidomimetic FTI-277 selectively blocks oncogenic RAS signaling by inducing cytoplasmic accumulation of inactive RAS-Raf complexes. J Biol Chem 1995; 45:26802-6.
148. Zhao J, Kung HF, Manne V. Farnesylation of p21 RAS proteins in *Xenopus* oocytes. Cell Mol Biol Res 1994; 40:313-21.
149. Sun J, Qian Y, Hamilton AD et al. RAS CAAX peptidomimetic FTI 276 selectively blocks tumor growth in nude mice of a human lung carcinoma with K-RAS mutation and p53 deletion. Canc Res 1995; 55:4243-7.
150. Kohl NE, Omer CA, Conner MW et al. Inhibition of farnesyltransferase induces regression of mammary and salivary carcinomas in RAS transgenic mice. Nature Med 1995; 1:792-7.
151. Yan N, Ricca C, Fletcher J et al. Farnesyltransferase inhibitors block the neurofibromatosis type I (NF1, malignant phenotype). Cancer Res 1995; 55:3569-75.
152. McCormick F. Activators and effectors of RAS p21 proteins. Curr Opin Genet Dev 1994; 4:71-6.
153. Lebowitz PF, Davide JP, Prendergast GC. Evidence that farnesyltransferase inhibitors suppress *RAS* transformation by interfering with Rho activity. Mol Cell Biol 1995; 15:6613-21.
154. Cox AD, Brtva TR, Lowe DG et al. R-RAS induces malignant, but not morphologic, transformation of NIH3T3 cells. Oncogene 1994; 9:3281-8.
155. Carboni JM, Yan N, Cox AD et al. Farnesyltransferase inhibitors are inhibitors of RAS but not R-RAS2/TC21 transformation. Oncogene 1995; 10:1905-13.
156. Dalton MB, Fantle KS, Bechtold HA et al. The farnesyl protein transferase inhibitor BZA-5B blocks farnesylation of nuclear lamins and p21ras but does not affect their function or localization. Cancer Res 1995; 55:3295-304.
157. Kawabata M, Imamura T, Miazono K et al. Interaction of the transforming growth factor-β type I receptor with franesyl-protein-transferase-α. J Biol Chem 1995; 270:29628-31.
158. Wang T, Danielson PD, Li B et al. The $p21^{RAS}$ farnesyltransferase α subunit in TGF-β and activin signaling. Science 1996; 271: 1120-2.

CHAPTER 5

Mammals II: Downstream of RAS and Actin-Cytoskeleton

Hiroshi Maruta

In the two preceding chapters we have learned how (i) RAS is translocated from the cytoplasm to the plasma membrane by its C-terminal farnesylation and a few other covalent modifications, and how (ii) the inactive GDP-bound form of RAS (D-RAS) on the plasma membrane is then converted to the active GTP-bound form (T-RAS) by SOS and a few other GDP/GTP exchange factors (GEFs). In this chapter we will discuss: (i) the role of several distinct RAS GTPase activating proteins (GAPs) in converting the normal T-RAS to D-RAS; (ii) the oncogenic mutants of RAS that are resistant to the action of GAPs; (iii) the role of several distinct effectors of T-RAS in transmitting the oncogenic (or mitogenic) RAS signals further downstream; (iv) the role of actin-cytoskeletal proteins in the regulation of RAS signaling; (v) three G proteins in the Rho family which are involved in the reorganization of the actin-cytoskeleton caused by T-RAS and (vi) the genes whose expression is upregulated or downregulated by T-RAS.

I. RAS GAPs

The discovery of RAS GAP activity in crude extracts of both *Xenopus* oocytes and mammalian cells by McCormick and his colleague in 1987[1] was perhaps one of the most critical and exciting moments in the history of RAS research. They found that the intrinsic low GTPase activity of normal RAS, but not the oncogenic RAS mutants, is highly stimulated by a protein(s), which they coined RAS GTPase activating

Regulation of the RAS Signaling Network, edited by Hiroshi Maruta and Antony W. Burgess.

proteins (GAPs), in the cell extracts. Their finding provided not only the basis for the mechanism underlying a rapid conversion of normal T-RAS to D-RAS, but also the clue to our understanding why, unlike the normal RAS, oncogenic RAS mutants continue to remain in the active form (T-RAS) for a prolonged time within living cells and therefore cause the development of tumors in various tissues of human and other animals. Gibbs and his colleagues purified and cloned the first Ras GAP of 120 kDa, called GAP1, from bovine brain.[2,3]

It consists of 1044 amino acids, and its GAP activity resides in its C-terminal domain of 325 amino acids (GAP1C, residues 720-1044).[4,5] GAP1C binds both normal T-RAS and oncogenic RAS mutants, but activates only the normal RAS GTPases to attenuate its normal signal transducing activity by converting the T-RAS to D-RAS, and then detaches from D-RAS.[3] Thus, GAP1C appears to serve as an attenuator of the normal T-RAS, but fails to attenuate the oncogenic RAS mutants. However, if we are able to design and create a specific mutant of GAP1C or other RAS GAPs that can stimulate the extremely low intrinsic GTPase activity of various oncogenic RAS mutants, this "putative" GAP mutant could serve as a potent anti-RAS cancer drug which would be potentially useful for the treatment of RAS-associated cancers. Such a "super" GAP is one of the "Holy Grails" of our anti-RAS cancer research. The N-terminal domain of GAP1 (GAP1N) contains at least five distinct functional domains (see Fig. 5.1): (i) the N-terminal Pro-rich motif that binds the SH3 domain of the Src Tyr kinase family;[6] (ii) the SH3 domain that binds a cytoplasmic protein of 68 kDa called p68 or G3BP;[7] (iii) two distinct SH2 domains which flank the SH3 domain, and bind at least two Tyr-phosphorylated pro-

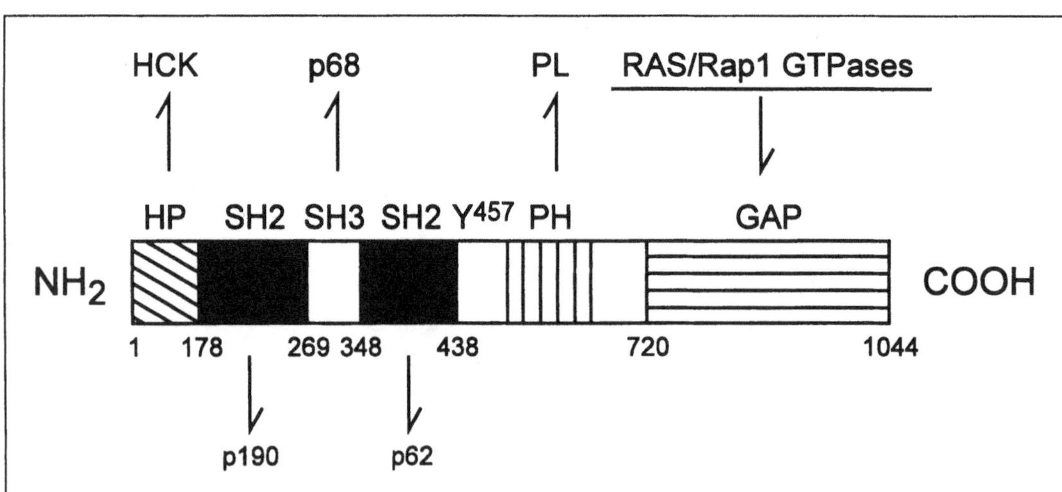

Fig. 5.1. Functional domains of GAP1 (Bovine).

teins called p190 and p62;[8] (iv) the PH (pleckstrin-homology) domain.[9] We will discuss the biological function of these N-terminal domains in detail later.

A second GAP of 28 kDa called GAP2 that was isolated shortly after the discovery of GAP1 is not a GAP for RAS, but for another small G protein (or GTPase) called Rho.[10] GAP2 stimulates the intrinsic GTPase activity of Rho, but not that of RAS or its cousin called Rap1. Sometime later the cDNA for GAP2 was cloned, and its sequence has revealed that the initially purified GAP2 of 28 kDa is just the C-terminal fragment of the full-length GAP2 of 50 kDa.[11] Besides GAP2, several other Rho GAPs were cloned.[12] One of them is p190, which binds SH2 domains of GAP1.[13] We will come back to these Rho GAPs later. A third GAP(s) of 55 or 88 kDa called GAP3c (c for cytosolic) or GAP3m (m for membrane-bound) that was isolated or cloned after the finding of GAP2 stimulates only the intrinsic GTPase activity of Rap1, but not those of RAS or Rho.[5,14,15] GAP1 is highly specific for RAS, and does not stimulate the intrinsic GTPase activity of Rho or Rap1. Thus, each of these GAPs has a very high specificity for the corresponding GTPase. We will discuss the residues of RAS, Rho and Rap1 that determine their GAP specificity later.

In 1990 two groups independently cloned a new tumor suppressor candidate gene product called NF1 or Neurofibromatosis type 1.[16,17] Deletion or point mutations of NF1 are associated with the development of the fibrous tumors in neuronal tissues, such as schwannomas. NF1 consists of 2818 amino acids, and its sequence has revealed that it contains a GAP1C-related domain called GRD.[17,18] In the NF1-GRD, we found a subdomain of 78 amino acids (NF78, residues 1441-1518) that is sufficient to stimulate the intrinsic GTPase activity of normal RAS, but not the oncogenic RAS mutants, as does GAP1C.[19] Although neither NF1-GRD nor NF78 is a "super" GAP that activates the oncogenic RAS GTPases, overexpression of NF1-GRD, NF78 or even a smaller NF1 fragment (NF56, residues 1441-1496) can suppress malignant transformation of NIH/3T3 cells caused by oncogenic RAS mutants such as v-Ha-RAS.[19,20] This observation confirmed that NF1 is indeed a tumor suppressor. Furthermore, the schwannoma cells that lack a functional NF1 show a much higher intracellular level of T-RAS (GTP-bound form of RAS) than the normal Schwann cells.[21,22] These observations suggest that NF1 plays a major role in the negative regulation (attenuation) of T-RAS in these cells. Although both GAP1 and NF1 bind Rap1 GTPase, both GAPs stimulate the intrinsic GTPase activity of only RAS, and not Rap1 or any other G proteins.[23,24]

RAS and Rap1 GTPases share overall 50% sequence identity, and in their so-called effector domains (switch I, residues 32 to 40) 100% sequence identity. Both GTPases require their switch I domains for their activation by the corresponding GAPs (GAP1 or NF1 for RAS, and GAP3c or GAP3m for Rap1). Why cannot RAS GAPs activate

Rap1 GTPases? Why cannot Rap1 GAPs activate RAS GTPases? In an attempt to identify the RAS and Rap1 residues that determine their GAP specificity, we have created a series of RAS/Rap1 chimeras and RAS or Rap1 mutants. In the end what we found is that their second effector (switch II) domains (residues 61-65) are solely responsible for their GAP specificity.[5,25-27] GTPases must contain both Gln61 and Ser65 to be activated by Ras GAPs (GAP1 and NF1), whereas they must have Thr65 to be activated by Rap1 GAPs (GAP3c and GAP3m). All of the remaining GTPase residues are functionally replaceable between RAS and Rap1 for activation by the corresponding GAPs. Furthermore, we found that both RAS and Rap1 GTPases must have Tyr, Phe or Leu at position 64 to be activated by GAPs. Thus, it appears that each GAP interacts with its target GTPase through two distinct GTPase domains, i.e., the common "switch I" domain and the unique "switch II" domain that determines the GAP specificity of each GTPase to activate the corresponding GTPases. Similarly, Rho GTPases also require both the "switch I" domain (residues 34-42) and "switch II" domain (residues 63-67) to be activated by a Rho GAP such as GAP2, but not by the C-terminal domain of p190 that also has a Rho GAP activity.[28] Replacement of Asp at either position 65 or 67, which correspond to positions 63 and 65 of RAS, by the RAS residues (Glu63 or Ser65) completely abolishes the GAP2-dependent activation of Rho GTPases.[28]

Besides GAP1 and NF1, at least two additional RAS GAPs have been purified or cloned (see Table 5.1). One is called GAP100, as it is a GAP of 100 kDa which is highly specific for RAS GTPases as are GAP1 and NF1.[29,30] The other is called GAP/IP4, as it is a unique GAP of 829 amino acids that binds inositol tetrakisphosphate (IP4), and stimulates the intrinsic GTPase activity of both normal RAS and Rap1.[31] GAP/IP4 is the first GAP that can activate both RAS and Rap1 GTPases. Interestingly IP4 stimulates its GAP activity towards

Table 5.1. Mammalian GTPase activating proteins (GAPs)

GAPs	GTPases	Refs.
GAP1 (120 kDa)	RAS (H, K, H, R)	2, 3
GAP2 (28/50 kDa)	Rho	10, 11
GAP3 (55/88 kDa)	Rap1 (A, B)	5, 14, 15
NF1	RAS (H, K, N, R)	16-19
p190	Rho, Rac, CDC42	12, 13
Myosin Myr5	Rho, Rac, CDC42	122
GAP100 (100 kDa)	RAS (H, K, N, R)	29, 30
GAP/IP4	RAS, Rap1	31
GAP-R/p98	R-RAS	146

RAS GTPases, but not towards Rap1 GTPases, whereas phospholipids inhibit only its RAS GAP activity, and not its Rap1 GAP activity,[31] suggesting the possibility that RAS and Rap1 GTPases are activated by two distinct domains of GAP/IP4. Again, neither GAP100 nor GAP/IP4 is a "super" GAP that activates oncogenic RAS GTPases. However, GAP/IP4 could potentially provide a hint as to how to design a "super" GAP. As I have mentioned previously, GAP1, NF1 and GAP/IP4 bind Rap1, but only GAP/IP4 can stimulate the intrinsic GTPase activity of Rap1 as GAP3 (c/m). So if one could find in their GAP domains some common short sequence or structure between GAP/IP4 and GAP3 (c/m), and if this sequence/structure differs between GAP/IP4 and GAP1/NF1, the insertion of a distinct but analogous sequence/structure into an appropriate position of a Ras GAP might convert it to a "super" GAP molecule that activates oncogenic RAS GTPases, instead of Rap1 GTPases.

As you have seen, each GTPase needs both "switch I" and "switch II" domains to be activated by the corresponding GAPs. Therefore, it should not be so unreasonable to assume that each Ras GAP also contains two separate GTPase-binding domains, one for its interaction with "switch I" and the other for its interaction with "switch II", to activate its target GTPases. Indeed, as we will discuss in detail later, the N-terminal half of Raf (an oncogenic RAS effector) contains two separate RAS binding domains called sites I (residues 51-131) and II (residues 152-184) which bind T-RAS independently.[32] Interestingly, although the Raf site I fragment (Raf81) does not have any detectable RAS GAP activity (Fridman M and Maruta H, unpublished observation), it was reported that the N-terminal half of Raf (residues 1-257) shows a weak RAS GAP activity,[33] suggesting that the site II of Raf is required for its GAP activity. The interaction of RAS with the site I, but not the site II, of Raf requires the "switch I" domain of RAS, whereas the interaction of RAS with the site II, but not the site I, of Raf requires Asn26, Val45 and Ala59 of RAS, which are located in two separate so-called "activation" domains flanking the "switch I" domain and are also required for the oncogenicity of T-RAS.[27,32,34] Thus, it appears that the site I of Raf is involved in its binding to the "switch I" of RAS, and the site II of Raf in its binding to the flanking "activation" domains of RAS.

II. ONCOGENIC MUTATIONS OF RAS

Before we discuss the several oncogenic effectors of T-RAS such as Raf and GAP1 in detail, I must summarize briefly a series of mutations that convert normal RAS to oncogenic molecules, and contribute to the development of around 30% of total human carcinomas, notably more than 90% of pancreatic and 50% of colon cancers.[35]

In human cells there are at least four distinct RAS molecules, called Ha-RAS, Ki-RAS, N-RAS and R-RAS.[36] So far only the first three

RAS molecules (Ha-RAS, Ki-RAS and N-RAS) have been shown to become highly oncogenic when they are mutated at one of the several "hot" positions: Gly12, Gly13, Ala59, Gln61 and Glu63. Interestingly, all of these oncogenic mutants of RAS show a significantly lower intrinsic GTPase activity than the wild-type RAS and, more importantly, their GTPase activity is no longer stimulated by any wild-type RAS GAPs so far isolated or cloned. Therefore, these oncogenic mutants of RAS are almost permanently locked in the active GTP-bound form.

Among these oncogenic RAS mutants, v-Ha-RAS was the first that was identified as a gene product of a murine retrovirus called Harvey Sarcoma virus.[36,37] v and Ha stand for viral and Harvey, respectively. It was later found that another retrovirus called Kirstin Sarcoma virus also expresses a similar oncogenic RAS molecule called v-Ki-RAS (Ki for Kirstin).[36,37] N-RAS was found in a human neuroblastoma cell line (N for neuron).[36,37] Eventually it was found that both v-Ha-RAS and v-Ki-RAS are oncogenic mutants of normal cellular RAS molecules called c-Ha-RAS and c-Ki-RAS (c for cellular), respectively.[36,37] In v-Ha-RAS Gly12 and Ala59 are replaced by Arg and Thr, respectively, whereas in v-Ki-RAS Gly12 is replaced by Val.[36,37] Besides these oncogenic mutations at positions 12 and 59, oncogenic mutations of Gln61 are often found in one of these three RAS molecules derived from human cancers. The typical one is a Leu61 mutation. Among the oncogenic RAS mutants, the Leu61 mutant of RAS shows the highest affinity for GAP1 and NF1.[38]

A further mutational analysis of oncogenic RAS mutants performed by Sigal and his colleagues revealed that mutations in the region between Tyr32 and Tyr40 abolish the oncogenicity of v-Ha-RAS, suggesting that this 9-amino acid domain is required for RAS to interact with its downstream effectors.[39] Since then this domain has been named the "effector" domain or "switch I" domain. Later we found a second "effector" domain of RAS (residues 61-65) in which Tyr64 is required for oncogenicity.[27] This 5-amino acid domain is then called the "switch II" domain.

In 1989 a RAS-related G protein called Rap1 (or Krev-1) was reported to suppress malignant transformation caused by v-Ki-RAS.[40] As I have mentioned before, Rap1 shares 50% sequence identity with RAS, and the "switch I" domains of Rap1 and RAS are identical.[41] Although Rap1 binds both GAP1 and NF1 much more tightly than RAS, its intrinsic GTPase activity is not stimulated by these RAS GAPs.[23,24] Instead, Rap1 GTPases are activated by GAP3 (c/m). Mutations of Rap1 such as Val12 and Thr59 that block GAP3-dependent activation of Rap1 GTPases and therefore lock Rap1 almost permanently in the active GTP-bound form, fail to convert Rap1 to an oncogenic molecule. On the contrary, it was reported that a Glu63 mutation, which greatly reduces the intrinsic GTPase activity of the wild-type Rap1,[5] potentiates its anti-RAS tumor suppressor activity.[42] Using a series of RAS/Rap1 chimeras and mutants of RAS in which RAS residues are replaced by the corresponding Rap1 residues, my group and others found that most RAS residues can be replaced by the corresponding Rap1 residues without any loss of oncogenicity, except for two RAS-specific domains (residues 26-31 and 41-54) which flank the "switch I" domain.[27,43] In particular Asn26, Glu31 and Val45 are absolutely essential for the oncogenicity of RAS.[27] These observations indicate that the "activation I" (residues 26-31) and "activation II" (residues 41-54) domains of RAS are involved in its interactions with a down-stream effector(s) such as Raf.

In fact, as I mentioned before, replacement in either the "activation I" or "activation II" domain of RAS by the corresponding Rap1 residues abolishes its binding to the site II of Raf in vitro, and the Ras-induced activation of the kinase Raf in vivo.[32] These observations suggest that Rap1 does not bind the site II of Raf and therefore fails to activate the kinase Raf, which is a key oncogenic effector, although it binds Raf only through the site I. For the oncogenicity of T-RAS, binding of both sites I and II of Raf is necessary. Thus, Rap1, which binds site I, but not site II, of Raf could form a "nonproductive" complex with Raf, and therefore competitively interfere with the formation of the "productive" complex between T-RAS and Raf. In other words, since the Raf site I fragment of 81 amino acids (Raf81) alone is also able to competitively interfere with the "productive" complex formation between T-RAS and full-length Raf, Raf81 could potentially be useful for genotherapy of RAS-associated cancers. In fact, we have recently demonstrated that overexpression of Raf81 can suppress v-Ha-RAS-induced malignant transformation of NIH/3T3 cells as does overexpression of RAS binding fragments of NF1, under conditions where overexpression of even the Glu63 mutant of Rap1 shows no detectable anti-RAS tumor suppressor activity.[19,20] Thus, it is also conceivable that the Raf site II fragment of 33 amino acids (Raf33) could suppress RAS-induced malignant transformation by interfering with the "productive" complex formation between T-RAS and full-length Raf. We are currently testing this notion, although the site I appears to bind T-RAS much more tightly than the site II.[32]

III. EFFECTORS OF RAS

Once RAS on the plasma membranes is activated, by SOS or a few other GEFs transiently or by oncogenic mutations constitutively, the RAS/GTP complex (T-RAS) activates a variety of effectors to transmit its mitogenic or oncogenic signal to further downstream targets before it binds GAPs for its attenuation.[44,45] So far several distinct RAS binding effectors have been identified and cloned from mammalian cells: Raf isozymes, PI-3 kinase, GAP1, PKC-ζ and RalGDS (Fig. 5.2).

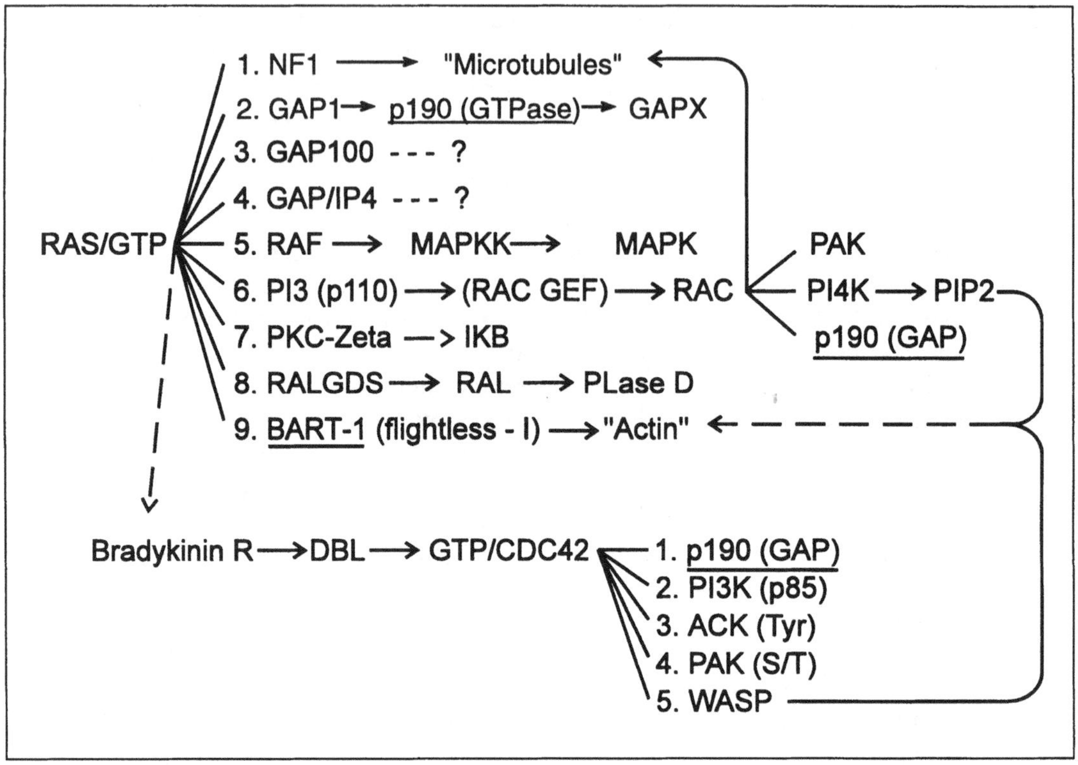

Fig 5.2. RAS signaling pathways.

Ser/Thr Kinase Raf

The three members of a Ser/Thr kinase family called Raf were among the first that were identified as mammalian RAS effectors: c-Raf-1, B-Raf/ REKS and A-Raf.[33,46-48] *Drosophila* Raf (D-Raf) is a homologue of mammalian B-Raf.[49] Like RAS (v-Ha-RAS or v-Ki-RAS), the first identified Raf molecule was v-Raf, an oncogenic mutant of the normal cellular c-Raf-1 integrated into a murine type C retrovirus called 3611-MSV, which is able to transform NIH/3T3 cells.[50] v-Raf lacks the N-terminal half of full-length c-Raf-1, but contains the C-terminal kinase domain.[50] Since a part of the missing N-terminal half serves as a negative regulator for the C-terminal kinase domain, v-Raf is a constitutively activated Ser/Thr kinase, whereas full-length c-Raf-1 is a latent kinase.[51] T-RAS is required for the activation of c-Raf-1, but v-Raf does not require T-Raf for its kinase activity and oncogenicity.[52] As discussed in detail in the following chapter, the major substrate for c-Raf-1 is MEK, a Tyr/Ser/Thr kinase that is activated by c-Raf-1 and in turn phosphorylates and activates another Ser/Thr kinase called MAP kinase.[53] It has been well established that the Raf-MEK-MAP kinase cascade plays a key role in T-RAS-induced signal transduction for mitogenesis or malignant transformation.[20,54,55] However, it still remains to be clarified how T-RAS activates Raf, as the direct interaction of T-RAS with Raf alone is not sufficient for the activation of normal full-length Raf. A third protein(s) must be involved in the T-RAS-induced Raf activation.

In this section I will focus the discussion mainly on the biological function of the N-terminal so-called "regulatory" domains of c-Raf-1 and a few other Raf isozymes (B-Raf and A-Raf). The N-terminal half (residues 1-305) of c-Raf-1 contains at least three distinct functional domains (see Fig. 5.3): (i) RAS binding site I (residues 51-131); (ii) RAS binding site II (residues 152-184); and (iii) the "negative regulator" domain (residues 274-305). As discussed in the preceding section, the site I of c-Raf-1 (Raf81), as well as the corresponding domains of B-Raf and A-Raf, tightly binds T-RAS[20,46-48] and over-expression of Raf81 alone is sufficient for suppressing v-Ha-RAS-induced malignant transformation.[20] These observations support the notions that (i) the site I fragment of c-Raf-1 alone could serve as a "dominant negative" mutant of c-Raf-1 and that (ii) c-Raf-1 acts as a key downstream effector of T-RAS for oncogenicity.

Recently two groups reported that full-length c-Raf-1 could be converted to an active kinase in a "RAS-independent" manner in vivo when it is attached to a RAS C-terminal CAAX motif, which allows the Raf to be farnesylated and translocated to the plasma membranes, suggesting that the major role of T-RAS on the plasma membranes in the activation of c-Raf-1 is simply to recruit the cytoplasmic c-Raf-1 to the plasma membranes.[56,57] However, the oncogenicity of this c-Raf-1/CAAX fusion protein turns out to be far weaker (2 or 3 orders of magnitude) than oncogenic RAS mutants,[58] indicating that (i) if the c-Raf-1/CAAX is fully activated, Raf alone appears to contribute only to a very small fraction (less than 1%) of the T-RAS-induced oncogenic

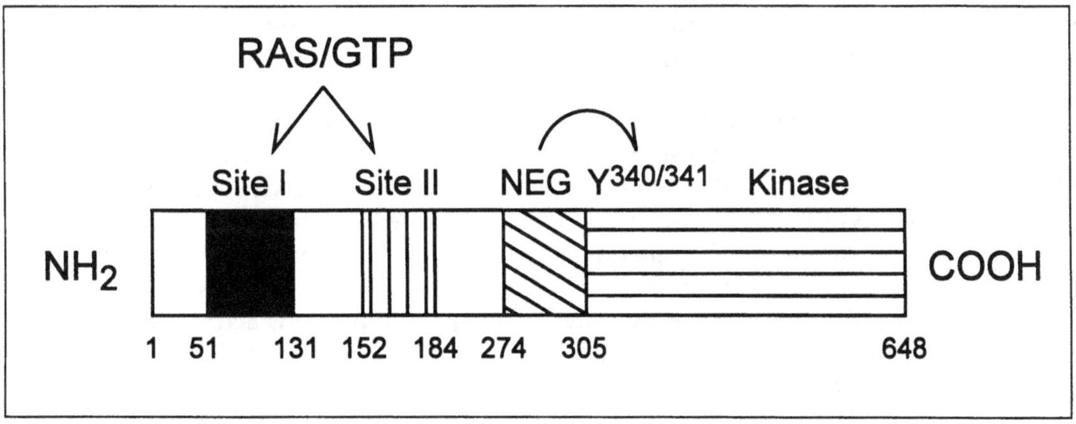

Fig. 5.3. Functional domains of RAF (C-Raf-1).

signaling, and that (ii) an additional effector(s) of T-RAS is absolutely required for the T-RAS-induced malignant transformation. In fact, although the oncogenicity of a constitutively activated mutant (Val12) of Rac1 alone is even weaker than that of Raf/CAAX, the Rac1 mutant greatly enhances the oncogenicity of the Raf/CAAX fusion protein.[58] Furthermore, a dominant negative mutant (Asn17) of Rac1 can suppress v-Ha-RAS-induced, but not the Raf/CAAX-induced, malignant transformation.[58] These observations clearly indicate that at least two independent signal transduction pathways induced by Rac1 and Raf are involved in T-RAS-induced malignant transformation.

Is CAAX attachment alone really sufficient for the full activation of Raf? If it were, since mutants of v-Ha-RAS which bind either the site I or site II of Raf are alone sufficient to recruit the Raf to the plasma membranes, these RAS mutants also could have activated Raf in vivo. However, all of the v-Ha-RAS mutants (Gly26 and Glu45, or His59 and Asn38) which bind only one of the two sites (I or II) on Raf fail to activate Raf in vivo,[32] and none of them is oncogenic at all—or their oncogenicity if any is too weak to be detected.[27] Furthermore, a constitutively activated Rap1 mutant which binds the site I, but not the site II, of c-Raf-1 also fails to activate the Raf, and is not oncogenic at all.[27,32] Perhaps one could argue that the Rap1 mutant is not oncogenic simply because it is associated with the Golgi apparatus instead of the plasma membranes.[59] However, a chimeric v-Ha-RAS/Rap1 which consists of the first 54 amino acids from v-Ha-RAS and the remaining residues from Rap1 is as highly oncogenic as the parental v-Ha-RAS, although it is geranylgeranylated (instead of farnesylated) and is therefore localized in the Golgi instead of the plasma membranes like Rap1.[5] Furthermore, *Drosophila* RAS1 which is normally geranylgeranylated is able to activate *Drosophila* Raf in vivo.[60] Thus, the reason why Rap1 fails to activate Raf is not simply its intracellular localization. Instead, it must be its failure in binding to the site II of Raf. Thus, it is most likely that T-Ras not only recruits cytoplasmic Raf to the membrane fractions (plasma membranes/Golgi apparatus), but also does something else for the activation of Raf, probably through its binding to the site II of Raf. Interestingly, a recently cloned Raf-related kinase called KSR (kesera) from human, *C. elegans* and *Drosophila* also contains in the N-terminal half a Cys-rich domain which, like the site II of Raf, consists of a highly conserved 6 Cys residues.[60] Genetic evidence indicates that KSR is required for the activation of Raf.[60] However, it still remains to be clarified biochemically whether KSR acts downstream of T-RAS to activate Raf, or whether KSR and T-RAS activate Raf independently.[60] Perhaps T-Ras might activate KSR by binding this Cys-rich domain, and in turn KSR activates Raf. Since Rap1 fails to bind the site II of Raf, it is most likely that Rap1 does not bind this Cys-rich domain of KSR. For details of KSR, see chapters 1 and 2.

A third functional domain of c-Raf-1, i.e., the so-called "negative regulator" domain, has been identified by monitoring the transforming ability of a series of N-terminal deletion mutants of the Raf.[61] As I mentioned earlier, full-length c-Raf-1 is not oncogenic at all. The deletion of the first 273 amino acids does not significantly increase the potency of its oncogenicity. However, a further deletion of the following 30 amino acids (residues 274-303) greatly increases its oncogenicity, and finally the deletion of the N-terminal 305 amino acids results in the maximum oncogenicity.[61] A further N-terminal deletion reduces its oncogenicity. These results clearly indicate that the "negative regulator" domain must reside between residues 274 and 305. However, it still remains to be clarified how this "negative regulator" domain of 32 amino acids blocks the kinase action of the C-terminal half.

Recently Morrison and her colleagues reported that replacement of the two major Tyr-phosphorylation sites of c-Raf-1, i.e., Tyr340 and Tyr341, by acidic amino acids such as Asp significantly stimulates the kinase activity of the Raf, and that cotransfection of v-Src and Raf in Sf9 insect cells results in phosphorylation of these two Tyr residues and stimulation of the Raf kinase activity.[62] These observations suggest that the phosphorylation of these two Tyr residues releases the possible intramolecular locking between them and the "negative regulator" domain. We are currently examining whether this "negative regulator" domain interacts directly with these two Tyr residues of the Raf.

Since the RAS binding site I on c-Raf-1 (Raf81, residues 51-131) was identified, and then shown to suppress v-Ha-RAS-induced malignant transformation,[20] a few groups including our own have created a series of Raf81 mutants in an attempt to identify the Raf residues that are critical for its binding to T-RAS, and also to screen a unique mutant(s) of Raf81 which shows a much higher affinity for T-RAS, and therefore would serve as a more potent anti-RAS tumor suppressor(s). Among the 81 amino acids in the site I, Arg89 is the first residue that was shown to be essential for RAS binding: replacement of this residue by Leu completely abolishes the RAS binding.[63] The corresponding mutation on *Drosophila* Raf, called D-Raf, significantly impairs the development of eyes, in particular their photo-receptor cell called R7, which requires the SOS/RAS/D-Raf/MEK signal transduction pathway.[64] We found that Arg89 in loop 3 cannot be replaced by any other residue, even Lys, without any loss of RAS binding (Fridman M, Treutlein H and Maruta H, submitted for publication). Furthermore, we found that at least four noncharged residues in the N-terminal half of Raf81 play a critical role in RAS binding (see Fig. 5.4): replacement of either Gln64 (loop 1), Val70 (β-sheet 2), Ala85 (α-helix 1) or Val88 (loop 3) by basic amino acids (Arg, Lys or His) greatly increases the affinity for T-RAS, but replacement of the same residues by acidic amino acids such as Asp almost completely abolishes RAS binding (Fridman M, Treutlein H and Maruta H, submitted for publication). Replacement of either Leu62 (β-sheet 1) or Leu91 (loop 3) by Arg also abolishes the RAS binding completely (Fridman M, Treutlein H and Maruta H, submitted for publication). Replacement of Lys84 (α-helix 1) by Leu reduces the RAS binding by 90% (Fridman M, Treutlein H and Maruta H, submitted for publication). Marshall and his colleagues also found independently by Ala scanning mutagenesis that replacement of Leu91 by Ala or Lys84 by Asp greatly reduces the RAS binding,[65] confirming that both residues are critical for the RAS binding.

The three-dimensional (3D) structure of both RAS and Rap1, as well as the Rap1-Raf81 complex, will be discussed in detail elsewhere by Wittinghofer and his colleagues. Since Raf81 binds both RAS and Rap1 through their switch I domains (residues 32-40), which are indistinguishable between RAS and Rap1, the interface structure of the Rap1-Raf81 complex is basically the same as that of the RAS-Raf81 complex. Thus, based on the 3D structure of the Rap1-Raf81 com-

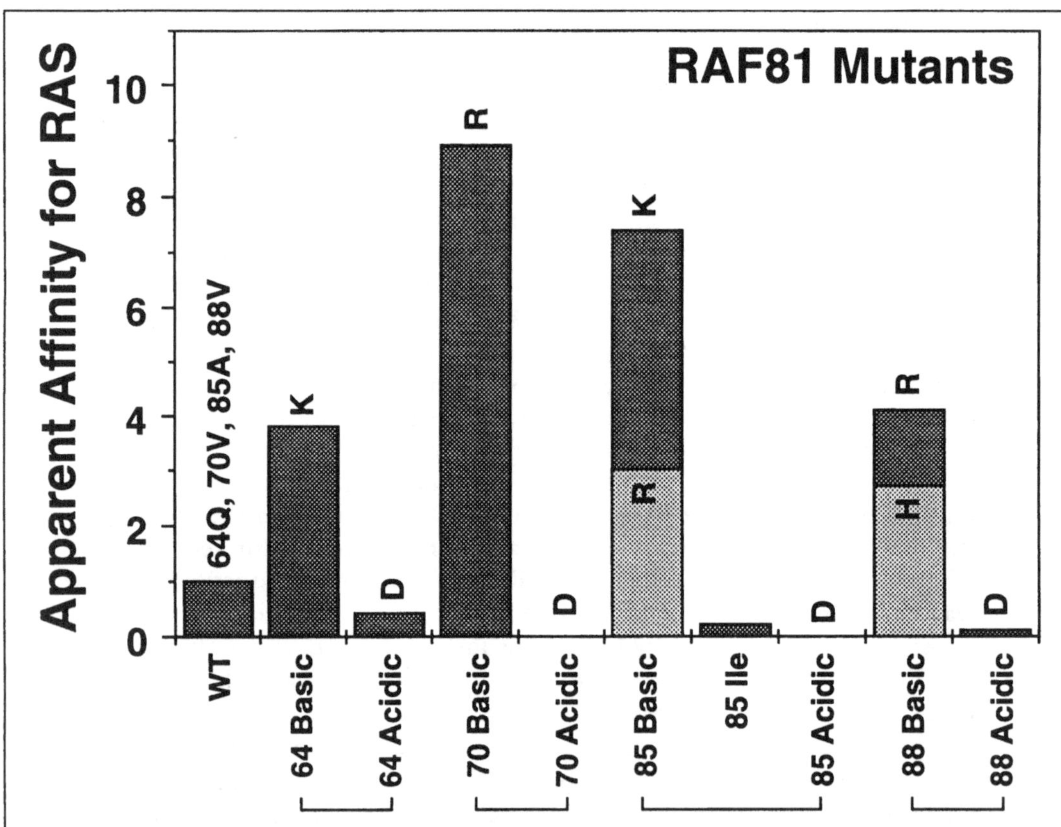

Fig. 5.4. Effects on RAS binding by replacements of four noncharged residues in the N-terminal half of Raf81.

plex,[66] I will discuss only briefly the possible role of a few key Raf residues in their interactions with the highly acidic switch I domain of RAS/Rap1 (see Fig. 5.5). First of all, Asp38 of RAS/Rap1, which directly binds Arg89 of Raf, we found cannot be replaced by any other amino acid without destroying the RAS-Raf81 interaction. This observation clearly explains why the RAS mutation at position 38 destroys its interaction with Raf81 and its oncogenicity. Although Asp33 of RAS/Rap1 does not directly bind Ala85 of Raf, these two residues are located very close to each other. Thus, it is possible that a Lys residue replacing Ala85 and greatly increasing the affinity of Raf for RAS could interact directly with Asp33 to form a second salt bridge. Our finding that replacement of Ala85 with acidic amino acids such as Asp abolishes RAS binding completely appears to support the above notion. The Ras switch I domain contains a third acidic amino acid at position 37 (Glu). Glu37 might form a third salt bridge with an Arg residue replacing either Val88 or Val70 of Raf and greatly increasing its affinity for RAS, as both Raf residues are located in the vicinity of Glu37, although Glu37 does not directly bind these Val residues. Our finding that replacement of these Val residues with acidic amino acids abolishes the RAS binding completely also appears to support the above notion. As I have mentioned, replacement of Asn64 of Raf by Lys, but not Asp, also greatly increases the affinity for RAS. Asn64 reportedly directly binds Arg41 just outside of RAS switch I domain.[66] However, the Asn64-Arg41 interaction is not critical for the Raf-RAS interaction, as replacement of Arg41 by Ala does not abolish either

	26 31	32	33 38	40 45
RAP	**gIFVEk**	**Y**	**DPTIED**	**SY RKQVe**
RAS	**NHFVDE---**	**Y**	**-- DPTIED--**	**SY--RKQVV**
		1	**: ::**	**1 11**
RAF	**SLHDCL**	**M**	**84KK-LKRR89**	**GL--QPECC**
		1	**:: ::**	**1**
PKC-Z	**GEDKSI**	**Y**	**RR-GARR**	**-WR-KLYRA**
GAP1	**EGVNPF**	**I**	**KS-NKHR**	**MI MFLDL**

Fig. 5.5. RAS-Raf (K85/R88) interaction.

the RAS-Raf interaction or the in vivo function of RAS in yeast and PC12 cells.[39,67] These observations suggest that the Lys64 mutation might allow Raf to form an additional salt bridge with an acidic residue of RAS, but outside of the switch I domain, perhaps at the highly conserved Asp57 which is reportedly required for the Raf-RAS interaction.[67]

Phosphatidylinositol (PI)-3 Kinase

PI-3 kinase is an enzyme that phosphorylates the D3 position of the inositol ring of PI and its phosphorylated derivatives to produce PI-3 phosphate [PI(3)P], PI 3,4-bisphosphate [PI(3,4)P_2] and PI 3,4,5-trisphosphate [PI(3,4,5)P_3]. This kinase consists of two distinct subunits called p110 and p85.[68] p110 is the catalytic subunit of 110 kDa, whereas p85 is the regulatory subunit of 85 kDa. p110 has to associate with p85 to be an active kinase. p85 contains several distinct functional domains (see Fig. 5.6): (i) an N-terminal SH3 domain (residues 10-78); (ii) a BCR domain (residues 129-254); (iii) two SH2 domains (residues 329-416 and 623-706) and (iv) a p110-binding domain (residues 425-600) called Inter-SH2 domain, that is flanked by the two SH2

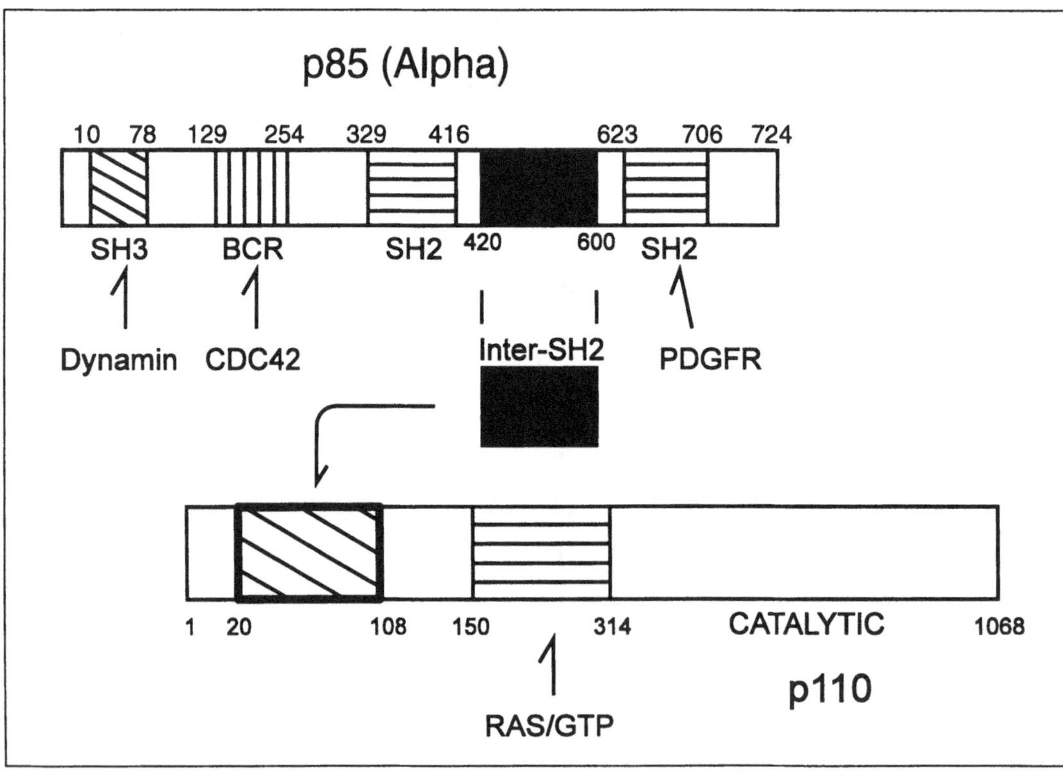

Fig. 5.6. PI3 Kinase: Subunits and functional domains.

domains. p110 contains at least three distinct functional domains: (i) an N-terminal p85-binding domain (residues 20-108); (ii) a RAS binding domain (residues 150-314) and (iii) a C-terminal catalytic domain.

PDGF or EGF stimulation of cells leads to the activation of RAS and PI-3 kinase.[44,45] However, it was not clear until recently whether RAS acts upstream of PI-3 kinase, or vice versa. A recent work by Downward's and Waterfield's groups established that T-RAS, but not D-RAS, binds p110 directly and activates PI-3 kinase.[69] What was not clear then is whether PI-3 kinase is required for the oncogenicity of T-RAS. To clarify this point, we created a dominant negative mutant of PI-3 kinase, i.e., the p110-binding fragment (Inter-SH2 domain) of p85, that blocks the p110-p85 complex formation, and therefore abolishes PDGF-induced activation of PI-3 kinase.[70] We found that the Inter-SH2 domain can suppress v-Ha-Ras-induced malignant transformation, clearly indicating that PI-3 kinase is absolutely essential for the oncogenicity of T-RAS.[70]

GAP1

Although GAP1 was initially discovered as an attenuator of normal RAS that converts T-RAS to D-RAS by stimulating its intrinsic GTPase activity, there is now increasing evidence indicating that it acts as a downstream effector of T-RAS as well. One of the earliest findings supporting this notion is that GAP1 blocks muscarinic K^+ channels of isolated atrial cell membranes in a T-Ras-dependent manner, and its effector activity resides in the N-terminal SH2/SH3 domains.[71] Recently Pawson's group found that disruption of GAP1 gene in mice causes both severe vascular system defects and neuronal apoptosis leading to embryonic lethality around day 10.[72] Since v-Ha-Ras transgenic mice grow normally and have never suffered from embryonic lethality, the major GAP1-null mutant phenotype is not due simply to the constitutive activation of normal RAS associated with the loss of the C-terminal GAP domain, but rather due to the loss of N-terminal "effector" domains of GAP1 (see Fig. 5.1). The same group also showed that overexpression of the N-terminal half of GAP1, containing the N-terminal Pro-rich motifs, two SH2 domains and an SH3 domain, causes a disruption of actin stress fibers as do oncogenic RAS mutants, but it is unable to transform normal NIH/3T3 cells.[73] However, we recently found that the deletion of the Pro-rich motifs allows the remaining N-terminal half of rat GAP1, in particular the SH2/SH3/SH2 domain (SH232, residues 164-449) alone, to transform rat 3Y-1 fibroblasts in the presence of a phorbol ester.[74] These observations altogether strongly suggest that (i) the SH232 domain of GAP1 acts as an oncogenic Ras effector, and that (ii) the N-terminal Pro-rich motifs (residues 1-163) of rat GAP1 are a negative regulator of the SH232 domain, and therefore could serve as a dominant negative mutant of the effector GAP1. Perhaps the ultimate proof for the effector role of GAP1 in oncogenic Ras signaling would be to demonstrate that overexpression of a GAP1 anti-sense RNA, blocking the de novo synthesis of GAP1, or its Pro-rich motifs can suppress malignant transformation caused by oncogenic Ras mutants.

How does the effector (SH232) domain of GAP1 exert its "oncogenic" action? The SH3 domain binds a cytoplasmic protein of 68 kDa called p68.[7] Overexpression of the SH3 domain alone, or microinjection of an antibody against the SH3 domain, blocks T-RAS-induced maturation of *Xenopus* oocytes,[75] indicating that (i) the interaction of the SH3 domain with an as yet unidentified protein target is essential for T-RAS signaling, and that (ii) the SH3 domain alone serves as a dominant negative mutant of the effector GAP1. In fact we have found that overexpression of the SH3 domain alone can suppress v-Ha-Ras-induced malignant transformation (He H and Maruta H, unpublished observation). Perhaps p68, a cytoplasmic target of the SH3 domain, might be involved in oncogenic Ras signaling. Interestingly, although p68 binds the SH3 domain alone without T-RAS, full-length GAP1 requires T-RAS for its binding to p68.[7] These observations indicate that some other domain(s) of GAP1 must block the p68-SH3 domain interaction in the absence of T-RAS, and T-RAS somehow releases this blocking. However, the N-terminal Pro-rich motifs do not bind the SH3 domain of GAP1, although it binds the SH3 domain of Hck or other members of the Src Tyr kinase family.[6]

The SH2 domains of GAP1 bind at least two distinct Tyr-phosphorylated proteins of 190 kDa and 62 kDa called p190 and p62.[8] There is so far no direct evidence indicating that p62 or its Tyr-phosphorylation is required for the oncogenicity of the SH232 or T-RAS. Interestingly, however, an antibiotic called radicicol/UCS1006, which suppresses T-RAS-induced malignant transformation, also inhibits Tyr-phosphorylation of p62, and therefore blocks its interaction with SH2 domains of GAP1.[76] Furthermore, in the presence of a phorbol ester that potentiates the oncogenicity of the SH232 domain of GAP1, the SH232 domain binds Tyr-phosphorylated p62 much more efficiently than full-length GAP1.[74] These observations suggest the possibility that Tyr-phosphorylation might be essential for malignant transformation caused by oncogenic RAS mutants. Nevertheless, so far nobody has shown that overexpression of p62 causes malignant transformation in normal cells, or that overexpression of a p62 anti-sense RNA suppresses T-RAS-induced malignant transformation. The interactions of GAP1 SH2 domains with the Tyr-phosphorylated forms of either p62 or p190 are independent of T-RAS.[77] Thus, it is also possible that neither p62 nor p190 is essential for the oncogenicity of T-RAS.

In fact, we found that p190 is not an oncogenic effector of either GAP1 or T-RAS, but instead p190 is a tumor suppressor that blocks oncogenic RAS signaling (Wang D, Nur-E-Kamal MSA, Tikoo A and Maruta H, submitted for publication). p190 contains at least two distinct functional domains:[13,78] (i) an N-terminal GTPase domain (residues 1-251), and (ii) a C-terminal GAP domain (residues 1177-1512). The extremely low GTPase activity of the p190 N-terminal domain is highly stimulated by an as yet uncharacterized GAP(s) present in an extract of mammalian cells.[78] This observation suggests that there must be a specific GDP/GTP exchange factor (GEF) which reconverts the inactive GDP-bound p190 (D-p190) to the active GTP-bound p190 (T-p190) (see Fig. 5.7). The C-terminal GAP domain stimulates the intrinsic GTPase activity of three members of the Rho GTPase family called Rho, Rac and CDC42, but not RAS.[13,28] Therefore, the C-terminal GAP domain could potentially serve as an attenuator of these G proteins of the Rho family. As discussed in a later section, it was recently shown that not only Rac but also Rho (and probably CDC42) are essential for oncogenic RAS signaling.[58,79-81] Thus, it is conceivable that p190 can suppress T-RAS-induced malignant transformation through its C-terminal GAP domain.

What we found recently was rather unexpected: the N-terminal GTPase domain of p190 alone is sufficient to suppress v-Ha-Ras-induced malignant transformation (Wang D, Nur-E-Kamal MSA, Tikoo

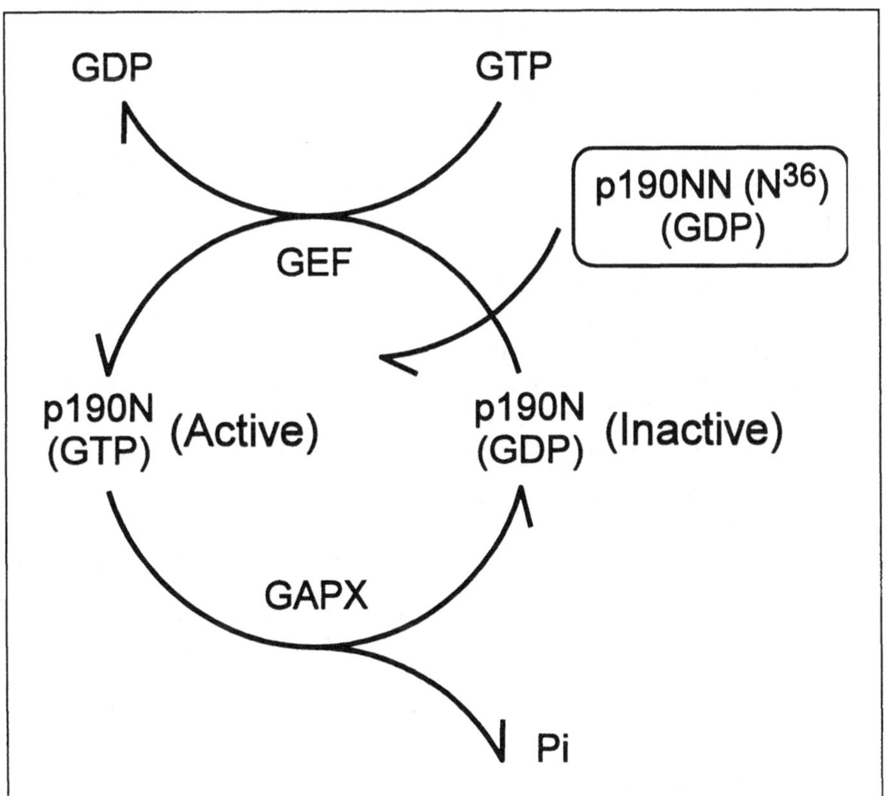

Fig. 5.7. GDP/GTP cycle of p190.

A and Maruta H, submitted for publication). Furthermore, we demonstrated that overexpression of a dominant negative mutant (Asn36) of the p190 GTPase domain (D-p190) alone or of a p190 anti-sense RNA is sufficient to transform normal NIH/3T3 cells, in particular in the presence of a phorbol ester. Interestingly, however, the oncogenicity of the p190 anti-sense RNA is much higher than that of the D-p190 mutant. These observations indicate that (i) not only full-length p190 but also its GTPase domain alone acts as a tumor suppressor, and that (ii) an as yet unidentified domain outside of the GTPase domain in p190 also contributes to the anti-oncogenicity of p190. In fact we found that the C-terminal GAP domain of p190 alone, which is an attenuator of Rho family G proteins, can suppress both RAS-induced malignant transformation and expression of the *c-Fos* gene (Nur-E-Kamal MSA et al, submitted for publication). Since c-Jun is required for oncogenicity of RAS, a complex of the two transcription factors c-Jun and c-Fos, called AP-1, is required for the activation of the *c-Jun* gene, and Rho/Rac/CDC42 are required for the activation of the *c-Fos* gene (81a), it is quite reasonable to conclude that the p190 GAP domain blocks oncogenic RAS signaling at least in part by blocking *c-Fos* gene expression.

How does the GTPase domain of p190 block the oncogenic action of T-RAS? Since this domain is a G protein (or GTP-dependent signal transducer), there must be a downstream target(s) which binds T-p190 (GTP/p190 complex), one of which could be a p190-specific GAP called GAPX. This target might activate an as yet uncharacterized signal transduction pathway which eventually blocks the oncogenic RAS signaling. Alternatively, the GTPase domain of p190 (p190-N) might suppress the oncogenicity of the effector GAP1 by binding to a pair of SH2 domains when p190-N is Tyr-phosphorylated (see Fig. 5.8). As I have mentioned before, the SH3 domain flanked by these two SH2 domains in GAP1 is essential for oncogenic RAS signaling. Thus it is possible that the "putative" p190-N/GAP1 interaction blocks the interaction of the GAP1 SH3 domain with its as yet unidentified "effector" protein(s). In fact, p190-N contains at least two Tyr residues which are phosphorylated by Src, although it remains to be clarified whether p190-N alone binds GAP1 (Roof R, personal communication). As the p68 requires T-RAS for its binding to the GAP1 SH3 domain, p68 could be an oncogenic effector of GAP1 or T-RAS. In this respect, it is of great interest to investigate whether overexpression of a p68 anti-sense RNA can suppress RAS-induced malignant transformation.

PKC-ζ

Most members of the protein kinase C (PKC) family are activated by phorbol esters. However, PKC-ζ is a unique isozyme of the PKC family which is not activated by phorbol ester.[82] Instead, PKC-ζ is

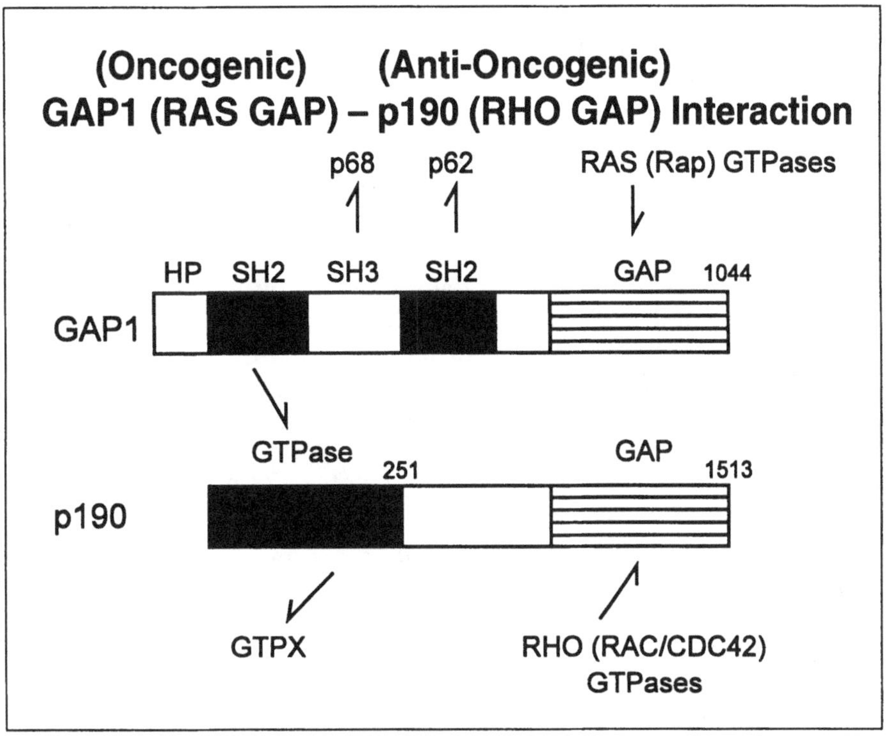

Fig. 5.8. Model for p190-N/GAP1 interaction.

activated by T-RAS.[83] It was reported that T-RAS, but not D-RAS, binds the N-terminal half (residues 1-194) of PKC-ζ directly.[84] The kinase activity of PKC-ζ resides in the C-terminal half, and this domain alone is a constitutively activated kinase.[85] Is PKC-ζ required for the oncogenicity of T-RAS? Yes, it is. It was reported that a few dominant negative mutants of PKC-ζ, which include a kinase-deficient mutant with or without the RAS binding domain, suppress malignant transformation caused by oncogenic RAS mutants.[85] These observations indicate that these dominant negative mutants of PKC-ζ block the RAS signaling, not by interfering with T-RAS/PKC-ζ interaction, but by interfering with the substrate/PKC-ζ interaction. Furthermore, these mutants of PKC-ζ do not affect v-Raf-induced malignant transformation, indicating that PKC-ζ and Raf are involved in two independent signal transduction pathways, both of which are activated by T-RAS.[84,85] It was also reported that a highly basic nona-peptide of PKC-ζ which corresponds to the pseudosubstrate motif (RRGARRWRK) within the N-terminal RAS binding domain blocks T-RAS-induced maturation of *Xenopus* oocytes.[83] This observation suggests that it is this nona-peptide motif that inhibits the action of the C-terminal kinase domain in full-length PKC-ζ. However, it still remains to be clarified how T-RAS is involved in the release of this intramolecular lock formed between the nona-peptide and the kinase domain.

RALGDS (RAL GDP-DISSOCIATION STIMULATOR)

RalA and RalB are two closely-related G proteins of 206 amino acids which share 50% sequence identity with RAS.[86] Human RalA shares 85% sequence identity with human RalB.[86] Like RAS, Ral is ubiquitously expressed across various tissues.[87] In 1993 a GDP/GTP exchange factor (GEF) specific for Ral was cloned by Weinberg's group,[88] and was called RalGDS (Ral GDP-Dissociation Stimulator). RalGDS is an ubiquitous phosphoprotein of 852 amino acids.[88] This protein is responsible for a rapid conversion of the inactive GDP-bound Ral (D-Ral) to the active GTP-bound Ral (T-Ral). In yeast two-hybrid systems RalGDS was found to be associated with T-RAS, but not D-RAS.[89,90] These observations suggest the possibility that T-RAS activates Ral through RalGDS. In fact T-RAS activates RalGDS.[91] Is Ral required for the oncogenicity of T-RAS? What is a down-stream effector of Ral? These two critical questions have been answered recently. One of the Ral effectors turns out to be phospholipase (PLase) D.[92] The first 11 amino acids of Ral are involved in the Ral-PLase D interaction.[92] Interestingly, v-Src stimulates PLase D activity in a T-RAS-dependent manner, and T-RAS also stimulates PLase D activity.[93] A dominant negative mutant of Ral inhibits v-Src/v-Ha-RAS-induced activation of PLase D, and a constitutively activated mutant (Leu72) of Ral enhances it.[92] Similarly, a dominant negative mutant of Ral suppresses v-Ha-Ras-induced malignant transformation (Brtva T et al, unpublished observation). These observations clearly indicate that RalGDS is an oncogenic effector of T-RAS. Interestingly the GTP-bound, but not the GDP-bound, form of Rap1 also binds RalGDS, as it binds any other known effectors of T-RAS.[89]

However, it still remains to be clarified whether PLase D is required for the oncogenicity of T-RAS, since an additional effector of T-Ral was cloned independently by three groups,[94-96] and was called Ral-binding protein 1 (RalBP1), Ral-interacting protein 76 (RLIP76) or Ral-interacting protein 1 (RIP1). This protein of 76 kDa contains two functional domains:[95] (i) a Ral-binding domain (residues 403-499), and (ii) a GAP domain (residues 210-353). The GAP domain stimulates the intrinsic GTPase activity of both CDC42 and Rac, but not Rho.[94-96] Thus, here we call this protein "RCR76" (for Ral/CDC42/Rac-binding protein of 76 kDa). T-RAS could link Rac/CDC42 indirectly, through these three proteins (RalGDS, Ral and RCR76). As I have mentioned before, T-RAS also links Rac/CDC42, but through two other proteins (GAP1 and p190). However, the exact physiological role or meaning of these two different links between T-RAS and Rac/CDC42 still remains to be clarified.

IV. ACTIN-CYTOSKELETON

In 1974 Weber and his colleagues found that a rapid disruption of actin stress fibers in fibroblasts or epithelial cells is an early intracellular event associated with their malignant transformation, caused by SV40 virus or other oncoviruses carrying RAS and Src oncogenes.[97] This finding triggered a series of biochemical studies attempting to explore the possible link between cancer and the actin-cytoskeleton. Since then more than 100 different actin-binding proteins, in particular a variety of cytoskeletal proteins localized at focal adhesion plaques (FAP), have been isolated, characterized or cloned.[98] The FAP is the site where cells adhere to the solid substratum through the interactions between the extracellular matrix proteins such as fibronectin and the transmembrane proteins such as integrins, and where actin stress fibers originate. Until recently, however, the molecular link(s) between RAS or other oncoproteins and the actin cytoskeleton remained quite obscure.

Actin-Cytoskeletal Proteins (ACPs) as Tumor Suppressors

The first breakthrough in this field was the finding during 1992-1993 that overexpression of genes encoding vinculin or α-actinin can suppress SV40-induced malignant transformation of NIH/3T3 cells.[99,100] Both proteins are F-actin cross-linking/bundling proteins associated with focal adhesion plaque (FAP). Interestingly, expression of several genes encoding these two actin-binding proteins and gelsolin is downregulated by malignant transformation caused by SV40 and RAS.[101]

Overexpression of a gelsolin mutant was also found to suppress Ras-induced malignant transformation.[102] Gelsolin is a Ca^{2+}-dependent F-actin capping/severing protein.[103] PIP_2 inhibits both F-actin capping and severing activities of gelsolin.[103] More recently we found that a few other actin-cytoskeletal proteins called tensin, NF2 and HS1 also suppress Ras-induced malignant transformation.[104-106] Tensin is an F-actin capping SH2 protein localized with FAP.[107] The C-terminal SH2 domain is sufficient for its FAP localization. Tensin contains three distinct F-actin binding domains (I, II and III) and domain III, located in the middle of the tensin molecule, is responsible for capping the barbed end of the actin filament. The N-terminal actin-binding domains I and II are responsible for crosslinking of actin filaments. However, it still remains to be clarified which domain of tensin is responsible for suppressing RAS-induced malignant transformation. NF2 is the neurofibromatosis 2 gene product (Merlin) whose N-terminal half is related to the corresponding domain of an F-actin capping protein called radixin.[104] We found that both the N-terminal half and the C-terminal half alone are sufficient for suppressing RAS-induced malignant transformation.[104,105] HS1 is an F-actin binding SH3 protein that is related to another F-actin binding SH3 protein called cortactin/EMS1.[106] These observations suggest that many other actin-cytoskeletal proteins (ACPs) also have an anti-RAS tumor suppressor potential (see Table 5.2), although the molecular mechanism underlying the anti-RAS action of some of these ACPs still remains to be determined.

Cytoskeletal SH3 Proteins

HS1 and EMS1 belong to a rapidly growing family of actin-cytoskeletal SH3 proteins.[108] Most of these proteins each contain a single SH3 domain at the C-terminus, except α-spectrin which contains an SH3 domain in the middle of the molecule (see Fig. 5.9). HS1 and EMS1 SH3 domains share 80% sequence identity, and both bind the Pro-rich motif of a single cytoplasmic protein of 45 kDa called Pash3.[106] Pash3 stands for a protein associated with SH3 domains (of HS1 and EMS1). We are currently examining whether Pash3 binds SH3 domains of two other cytoskeletal proteins called H74 and myosin IB, which share 60% and 50% sequence identity, respectively, with the SH3 domain of HS1. H74 is a human homologue of yeast CDC15 which becomes associated with the actin contractile ring during cytokinesis.[109] Both HS1 and EMS1 bind F-actin through the unique 37-amino acid motifs in their N-terminal halves (see Fig. 5.10). Full-length HS1 is localized in the cytoplasm, and suppresses v-Ha-RAS-induced malignant transformation of NIH/3T3 cells.[106] However, deletion of the C-terminal SH3 domain translocates the rest of the HS1 molecule to the nucleus, and abolishes its anti-oncogenicity.[106] These observations indicate that the SH3 domain is required for both the cytoplasmic localization and the anti-RAS action of HS1. HS1 contains the nuclear localization signal (NLS) between residues 263 and 274 (Fig. 5.10). The SH3 domain appears to mask the function of this NLS. Deletion of both the SH3 domain and the NLS brings HS1

Table 5.2. Tumor suppressor family of actin-cytoskeletal proteins

Proteins	Oncogenes	Refs.
1. Vinculin	SV40	99
2. α-Actinin	SV40	100
3. Gelsolin	H-RAS	102
4. NF2/Merlin [a]	H-RAS	104
5. Tensin	H-RAS	105
6. Cofilin [b]	H-RAS	105
7. HS1 [c]	H-RAS	106

[a] Related to TALIN, EZRIN, RADIXIN, MOESIN and Protein 4.1
[b] Related to GELSOLIN (F-Actin Severing/Capping, and PIP_2-Binding)
[c] Related to EMS1/CORTACTIN (F-Actin Binding SH3 Protein)

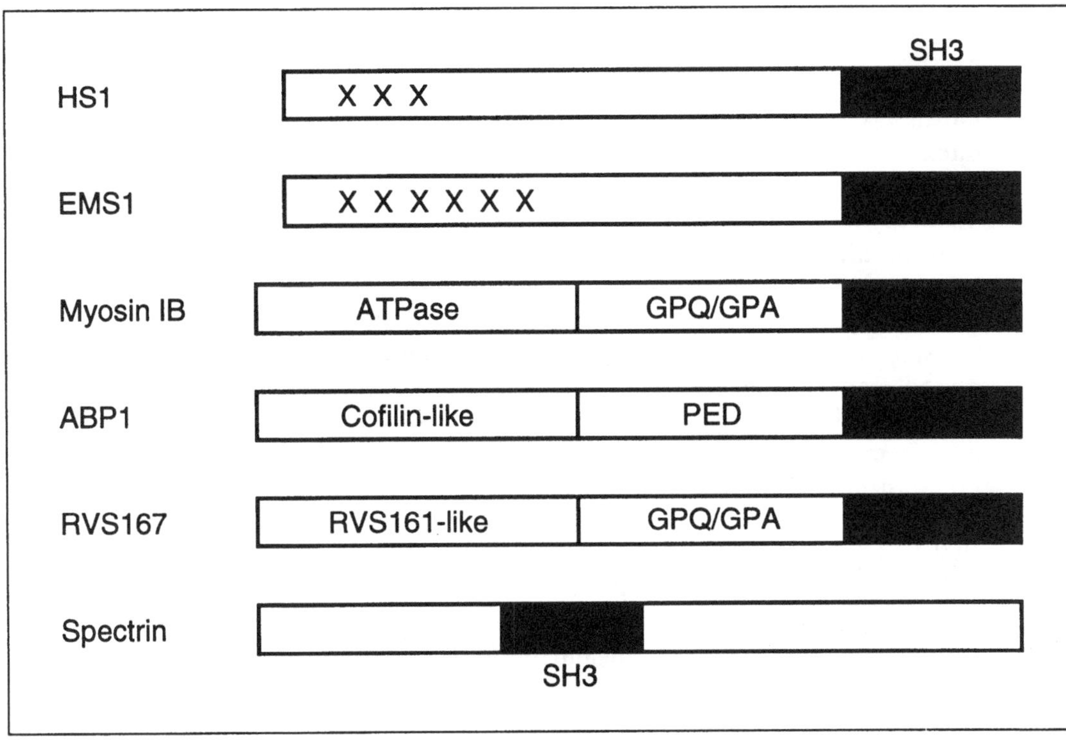

Fig. 5.9. Cytoskeletal SH3 proteins.

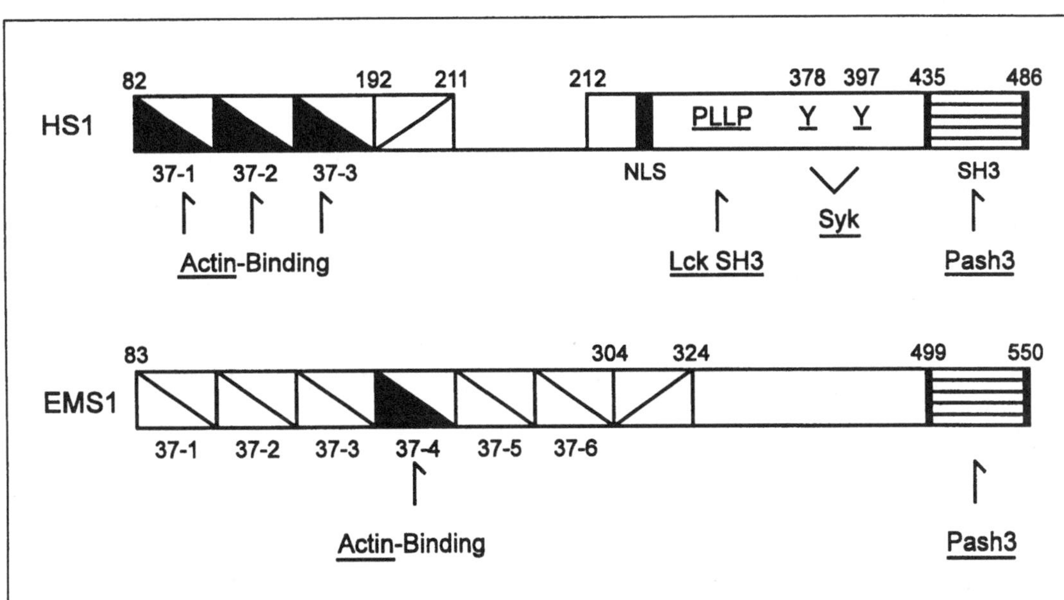

Fig. 5.10. HS1 and EMS1 (cortactin).

back to the cytoplasm, but fails to restore its anti-oncogenicity.[106] In addition to the SH3 domain, HS1 requires two critical phosphorylatable Tyr residues at positions 378 and 397, and the N-terminal actin-binding motifs for the anti-oncogenicity. Either replacement of these two residues by Asp or deletion of the actin-binding motifs (residues 82-192) completely abolishes the anti-RAS action, but still allows HS1 to remain in the cytoplasm.[106] Thus, either these two Tyr residues or the actin-binding motifs are not required for the cytoplasmic localization. However, when HS1 is Tyr-phosphorylated in B lymphocytes upon the activation of cell surface antigen receptor (BCR), HS1 is rapidly translocated into the nucleus.[110] This observation suggests that phosphorylation of some other Tyr residues in HS1 might be responsible for the nuclear localization. Interestingly, EMS1 lacks a nuclear localization motif corresponding to the HS1 NLS, and therefore EMS1 is not translocated to the nucleus even if it is Tyr-phosphorylated.

Flightless-I/Bart-1

Last year a novel protein called flightless-I (eye) or Bart-1 (one) was shown to bind both actin and T-RAS.[111,112] Bart-1 stands for the first protein that brings actin and RAS together. This finding would be another breakthrough for our better understanding of the link between T-RAS and actin-cytoskeleton. The N-terminal half of human Bart-1, which is related to the Leu-rich repeats (LRR) of yeast adenylate cyclase (YAC), binds T-RAS but not D-RAS.[111] T-RAS activates YAC directly.[113] The C-terminal half of Bart-1 is related to gelsolin and binds actin directly.[112] Since overexpression of full-length Bart-1 suppresses T-RAS-induced heat-shock sensitivity in yeast,[111] it is most likely that Bart-1 is a tumor suppressor that blocks the oncogenic action of T-RAS. We are currently overexpressing the RAS binding or actin-binding domains of Bart-1 in RAS-transformants of NIH/3T3 cells, to confirm its anti-RAS tumor suppressor activity and further identify the specific domain(s) that are responsible for the possible anti-RAS action. A loss-of-function mutation of Bart-1 causes a flightless phenotype in *Drosophila*.[114]

Deletion of one copy of the human Bart-1 gene is associated with Smith-Magenis syndrome, a behavior disorder.[115] However, whether this disorder is caused by the development of a tumor in central nervous system (CNS) or muscle tissues still remains to be clarified.

V. G Proteins in the Rho Family (Rho, Rac and CDC42)

There is increasing evidence indicating that T-RAS requires three distinct G proteins in the Rho family, i.e., Rho, Rac and CDC42, for its oncogenicity. First of all, as I have mentioned before, a dominant negative mutant (Asn17) of Rac suppresses T-RAS-induced malignant transformation.[58] Furthermore, it inhibits T-RAS/v-Src-induced activation of c-Jun N-terminal kinases (JNKs), and T-RAS-induced membrane ruffling.[80,81,116] Second, RAS-transformation causes the upregulation of bradykinin-receptor expression[117] and consequently activates DBL, which is a GEF specific for CDC42.[118] Like T-RAS, DBL is oncogenic, activates JNKs and induces membrane ruffling; a dominant negative mutant (Asn17) of CDC42 inhibits the DBL-induced JNKs activation and membrane ruffling.[80,81,118,119] Third, a dominant negative mutant (Asn19) of RhoA, or a farnesylation inhibitor (L-739,749) which blocks the C-terminal modification of RhoB, inhibits T-RAS-induced malignant transformation.[79,120] However, unlike Rac and CDC42, Rho is not involved in either the activation of JNKs or membrane ruffling.[80,81,119] Instead, Rho is involved in the organization of actin stress fibers.[121] In the case of Rac and Rho, it still remains to be clarified how T-RAS activates these two G proteins. Since it is suggested that T-RAS-induced activation of Rac requires PI-3 kinase,[58] it is most likely that Rac acts downstream of PI-3 kinase.

The next question is how these three G proteins of the Rho family control the organization of actin-cytoskeleton. So far there is no evidence indicating that these three G proteins bind actin filament (F-actin) or monomer (G-actin) directly. However, it was shown recently that a unique single-headed myosin heavy chain called Myr5 (for the fifth unconventional myosin from rat) contains a GAP domain near the C-terminus which stimulates the intrinsic GTPase activity of Rho, CDC42 and Rac[122] as does the C-terminal GAP domain of p190.[13] In other words Myr5 is the first myosin that brings actin and Rho family directly together, as Bart-1 is the first protein that brings actin and RAS together.[111,112] Since the GAP domain of Myr5 can attenuate the GTP-dependent signal transducing activity of these Rho family G proteins which are required for the oncogenicity of T-RAS, Myr5 potentially could serve as a tumor suppressor that blocks the oncogenic RAS signaling as p190 does. Thus, it is of great interest to examine whether overexpression of a Myr5 anti-sense RNA causes a malignant transformation of normal cells.

Myr5 contains four IQ motifs, in addition to an N-terminal ATPase head and C-terminal GAP domain, which are thought to bind calmodulin-like light chains.[122] This finding suggests the possibility that the interaction of Myr5 with F-actin is regulated by both Ca^{2+} and calmodulin (or a related Ca^{2+}-binding protein), as is that of a brush border myosin I ATPase.[123] Nevertheless, it still remains to be clarified whether Myr5 plays any significant role in the Rho family-mediated regulation of actin-cytoskeleton organization.

Hartwig and his colleagues have recently shown that T-Rac uncaps the fast-growing (barbed) end of the actin filament (F-actin) by activating PI-4 kinase, which catalyzes the synthesis of phosphatidylinositol 4,5-bisphosphate (PIP_2) in platelets, but T-Rho does not.[124] Interestingly, the PIP_2-binding fragment of gelsolin inhibits T-Rac-induced PIP_2 synthesis and F-actin uncapping.[124] These observations strongly suggest that the gelsolin mutant (His321) suppresses T-RAS-induced malignant transformation by inhibiting the ability of T-Rac to uncap the barbed end of F-actin through its PIP_2-binding site, as well as its F-actin capping site. In support of this notion, we have found recently that a double mutant (Gln112/Gln114) of cofilin, a gelsolin-related F-actin severing/PIP_2-binding protein of 166 amino acids, which no longer binds F-actin but still binds PIP_2,[125] is able to suppress T-RAS-induced malignant transformation, and that tensin, which caps the barbed end of F-actin but does not bind PIP_2, can also suppress T-RAS-induced malignant transformation.[105] Thus, it is conceivable that RAS-associated cancers can be treated by cytochalasins, the unique antibiotics that cap the barbed end of F-actin and block membrane ruffling.[126] Similarly, if a new drug(s) is found that blocks the T-Rac-induced activation of PI-4 kinase or binds PIP_2 selectively, the drug would be potentially useful for the treatment of RAS-associated cancers.

VI. RAS-ACTIVATED OR -REPRESSED GENES

I have discussed several distinct molecular events caused by T-RAS in the cytoplasm. In this section I will discuss very briefly the T-RAS-induced events in the nucleus which have not yet been well defined at the molecular level. T-RAS exerts its oncogenic action through eventually activating or inactivating a variety of genes. As I mentioned earlier, the bradykinin receptor gene is one of the RAS-activatable genes,[117] whereas several distinct genes encoding actin-binding proteins such as gelsolin are among the RAS-suppressible genes.[101] Overexpression of some RAS-suppressible genes such as HOX 1.4 can suppress RAS-induced malignant transformation.[127] The HOX 1.4 gene encodes a transcription factor containing a homeodomain, and its expression can be highly activated by retinoic acid.[127] Thus, it is worth while to explore as many anti-RAS tumor suppressor genes as possible among these RAS-suppressible genes. Also, once a strong enhancer element in the promoter region of the RAS-activatable genes is identified, this RAS-responsive element could be linked to a potent anti-RAS tumor suppressor gene, to express this tumor suppressor gene in a highly selective manner in the RAS-transformed cells, for gene therapy of RAS-associated cancer.

DOWNREGULATED GENES

Table 5.3 shows a short list of the RAS-suppressible (downregulated) genes so far identified. Unfortunately, however, the exact molecular mechanism underlying the RAS-induced suppression of these genes has not yet been clarified, and is so far only poorly understood. Since a mutant of gelsolin is a potent anti-RAS tumor suppressor,[102] and expression of the gelsolin gene is highly upregulated by a specific group of antibiotics including trichostatin A (TSA) and its derivative trapoxin

Table 5.3. RAS-suppressible genes

1. Genes encoding actin-binding proteins:
   Gelsolin,[101] α-actinin,[101] α-actin,[101] Tropo-myosin-1,[101] Myosin light chain[147]
2. Genes encoding transcription factors:
   Hox 1.4,[127] MyoD1,[148,149] Thyroid-specific trans-acting factor (TgTF1)(150)
3. Genes encoding enzymes:
   Lysyl oxidase (RRG)[152]
4. Genes encoding cell surface receptors:
   Thy-1[153]
5. Genes encoding extracellular matrix proteins:
   Fibronectin,[151] Collagen Type III[151]
6. Genes encoding milk proteins:
   Whey acidic protein[161]
7. Genes encoding membrane-associated proteins:
   H-Rev107[162]
8. Miscellaneous Genes:
   322,[154] IAP (Intracisternal A Particle)[151]

or TPX,[128] it may be worth while to discuss the action of these drugs in detail. TSA and related drugs are both specific and very potent inhibitors of histone deacetylase.[128] At nanomolar concentrations both TSA and TPX (molecular masses around 300) block this enzyme, resulting in a rapid accumulation of highly acetylated histones in target cells. Histone acetylation then causes a change(s) in the structure of chromosomes and leads to activation of some genes such as the gelsolin gene. TSA binds the enzyme noncovalently and therefore its effect remains reversible, whereas TPX binds the enzyme covalently and its effect becomes irreversible. Both drugs are cell-cycle blockers that arrest cell growth at G_1 or G_2 phases. TSA causes a morphological reversion (cell flattening) of both Sis- and RAS-transformed NIH/3T3 cells at around 300 nM, which is accompanied by the reappearance of actin stress fibers.[129,130] We found recently that 30 nM of TSA is sufficient to suppress a malignant phenotype (anchorage-independent growth in soft agar) of v-Ha-RAS-transformed NIH/3T3 cells almost completely, and 100 nM of TSA is sufficient to block anchorage-dependent growth in a liquid culture of the RAS-transformants completely though in a reversible manner (Tikoo A, Yoshida M and Maruta H, unpublished data). Both TSA and TPX are a million times more effective than n-Butyrate, another inhibitor of histone deacetylase, for morphological reversion of these transformed cells.[128] The only problem in using either TSA or TPX for the treatment of RAS-associated cancers is that they are cytostatic for normal cells as well, although they do not cause any morphological changes.[128] Therefore, these potential anti-cancer drugs have to be conjugated with a ligand(s) specific for the cell surface receptor on RAS-transformed cells to acquire the desired highly selective toxicity against the cancer cells.

Another potentially interesting RAS-suppressible gene is H-Rev107, which encodes a membrane-associated cytoplasmic protein of 160 amino acids.[162] This ubiquitous protein is highly abundant in brain cells and fibroblast cell lines such as REF52 and EK-3 which are resistant to RAS-induced malignant transformation, but is hardly detectable in RAS-transformable cell lines such as NIH/3T3 and various tumor cell lines derived from pancreatic or colon carcinomas. Overexpression of this protein is highly toxic for RAS-transformed cells, and causes growth arrest or cell death. However, it is not toxic to normal cells at all. Interestingly, the expression of this gene is downregulated by a few oncogenes such as adenovirus E1a, SV40 large T and Myc that render REF52 or other RAS-resistant cells highly RAS-transformable.[162] These observations strongly suggest the possibility that H-Rev107 is one of the key elements responsible for RAS-induced growth arrest. Thus, it is of great interest to examine whether PC12 cells, which are normally induced by oncogenic RAS mutants to both growth arrest and neuronal differentiation instead of proliferation,[45] can be converted by overexpression of an anti-sense H-Rev107 RNA to highly RAS-transformable cells. Furthermore, if one could find a specific drug that upregulates the H-Rev107 gene, it would be potentially useful for the treatment of RAS-associated cancers.

The smooth muscle α-actin gene is a third RAS-suppressible reporter gene which has great clinical potential.[101,163] This reporter gene was reportedly derepressed in RAS-transformed cells by a novel quinoline derivative called SCH51344, and this drug reverses the malignant phenotype of various RAS-associated tumor cells.[163] Furthermore, the same drug can suppress the oncogenic action of several other oncoproteins such as v-Abl, v-Raf, v-Mos and a constitutively activated mutant of MEK.[163] However, it does not affect the kinase cascade leading to the activation of MAP kinase, indicating that its targets act downstream of MAP kinase or in an as yet uncharacterized parallel pathway.[163] As a great surprise, this drug does not inhibit NGF-induced neuronal differentiation of PC12 cells, which also requires RAS.[163] However, if the H-Rev107 gene is also upregulated by this drug, and it is this gene that is responsible for RAS-induced neuronal differentiation, the unusual response of PC12 cells to both RAS and this drug could be easily explained.

Upregulated Genes

Table 5.4 shows a short list of RAS-inducible (upregulated) genes so far identified. Here, from a medical aspect, I will discuss only a

Table 5.4. RAS-inducible genes

1. Genes encoding EGF-related cytokines: HB-EGF,[131] TGF-α,[155] Amphiregulin[156]
2. Genes encoding other cytokines: Vascular Endothelial Growth Factor,[136] Mob-1[157]
3. Genes encoding cell surface receptors: Bradykinin receptor[117]
4. Genes encoding cell cycle regulators: Cyclin D1[158]
5. Genes containing AP1-binding site: c-Fos, c-Jun[159]
6. Multidrug resistant genes: MDR1[160]

few genes encoding extracellular molecules such as the bradykinin receptor (BKR), HB-EGF which is a unique heparin-binding (HB) EGF analogue,[131] and an EGF antagonist called Argos.[132] For the other RAS-regulated genes, you may find a more detailed and comprehensive review(s) in several years. When normal Rat-1 fibroblasts are transformed by oncogenic RAS mutants, but not by Src, Sis or Myc, they become highly sensitive to the mitogenic action of bradykinin, which induces phosphatidylinositol turnover.[117] It turns out that T-RAS upregulates the expression of the BKR gene around 50 times.[117] As we have mentioned earlier, BKR activates DBL, which in turn activates CDC42, which eventually induces both membrane ruffling and activation of JNKs, which are involved in RAS-induced malignant transformation.[80,81,118,119] Since DBL is oncogenic, and the action of CDC42 appears to be required for the oncogenicity of T-RAS, bradykinin antagonists (BKAs) such as des-Arg9,[Leu8]-bradykinin alone could potentially suppress RAS-induced malignancy by blocking the activation of the overexpressed BKR. Furthermore, each BKA could serve as a highly selective ligand towards RAS-transformed cells when it is conjugated with other potential anti-RAS cancer drugs such TSA and TPX which upregulate the gelsolin gene, the farnesylation inhibitors such as L-739,749 which block the essential action of RhoB,[79] or a bacterial exoenzyme called C3 which inactivates Rho by ADP-ribosylating it at Asn41.[133]

Expression of several distinct EGF-related growth factors as well as VEGF (vascular endothelial growth factor) has been shown to be upregulated by T-RAS.[134-136] Since, like EGF, these EGF-related ligands such as TGF-α, HB-EGF and Amphiregulin (AR) can activate the EFG receptor (EGFR), which eventually activates normal RAS,[134] there is a possibility not yet proved that this autocrine loop may be essential or critical for the oncogenicity of T-RAS. However, it was shown recently by disruption of the TGF-α gene in mice that keratinocytes lacking TGF-α can still be transformed by oncogenic RAS mutants,[137] indicating that at least TGF-α is not absolutely required for the oncogenicity of T-RAS. In an attempt to determine whether EGFR expression is required for the oncogenicity of T-RAS, we are currently examining whether an EGFR anti-sense RNA can suppress RAS-induced malignant transformation. If T-RAS turns out to require this autocrine loop, some proper EGF antagonists could be used for the treatment of RAS-associated cancers.

HB-EGF, in particular that of the human, turns out to have a very unique property which is not shared by any other RAS-inducible EGF-related ligands:[138] HB-EGF is a high-affinity receptor for diphtheria toxin (DT). Furthermore DT blocks the interaction between EGFR and HB-EGF.[139] Thus, when human fibroblasts are transformed by oncogenic RAS mutants, they become highly sensitive to the toxic action of DT, which is exerted by ADP-ribosylation of protein elongation factor-2.[135] The RAS-transformants are 50-fold more sensitive to DT-mediated killing than normal cells.[135] The HB-EGF precursor is a transmembrane protein which serves as a cell surface DT receptor, and later during its maturation is cleaved off from the membrane in a PKC-dependent process.[140] Thus, DT or its derivatives could be potentially useful for the treatment of RAS-associated cancers. Interestingly, back in 1973, an Italian group demonstrated the inhibition of Ehrlich tumors (ascites or solid) by DT even in the mouse,[141] which is least sensitive to DT among mammals, simply due to the lowest affinity of mouse HB-EGF for DT.[138]

Drosophila contains at least two distinct EGF related ligands called Spitz and Argos which regulate the development of this fruit fly in opposite manners.[132] In 1988 Nuesslein-Vorhard's group first identified a *Drosophila* gene locus called *Spitz* which is involved in dorsal-ventral axis formation and neurogenesis.[142] Subsequent cloning of this gene revealed that the *Spitz* gene encodes an EGF-like growth factor of 203 amino acids.[143] Shortly thereafter Rubin's and other groups independently cloned another *Drosophila* gene called *Argos* or *Gil* (for giant lens), which encodes another EGF-like secreted protein of 420 amino acids that is required for viability and negatively regulates the determination of eye cells.[144,145]

Recently Freeman's group found that Spitz indeed activates *Drosophila* EGFR (D-EGFR), whereas Argos blocks the activation of D-EGFR, indicating that Argos is an antagonist of Spitz. It still remains to be clarified whether Argos and Spitz compete for binding to D-EGFR or Argos activates its own receptor, instead of D-EGFR, and in turn this putative receptor somehow inactivates D-EGFR.[132] Interestingly both Spitz and Argos conserve the six Cys residues within their EGF motifs which form the three disulfide bonds critical for EGFR-binding.[144,145] We are currently examining whether the *Drosophila* Argos can block the activation of mammalian EGFR; if it does, we intend to humanize Argos by either cloning a human homologue, if any, of the Argos or alternatively by creating a chimeric Argos/human EGF (or TGF-α) molecule which can block the activation of mammalian EGFR. Interestingly sequence search of several DNA sequence banks has revealed that the human genome contains sequences encoding several EGF-like proteins, one of which (EST1628/Harp) has Argos-like characteristics (see Fig. 5.11) (Burgess AW, unpublished data). We believe these "humanized" Argos molecules could be potentially useful for the treatment of cancers caused by overexpression of EGFR or perhaps oncogenic RAS mutants as well (if EGFR is required for the oncogenicity of RAS).

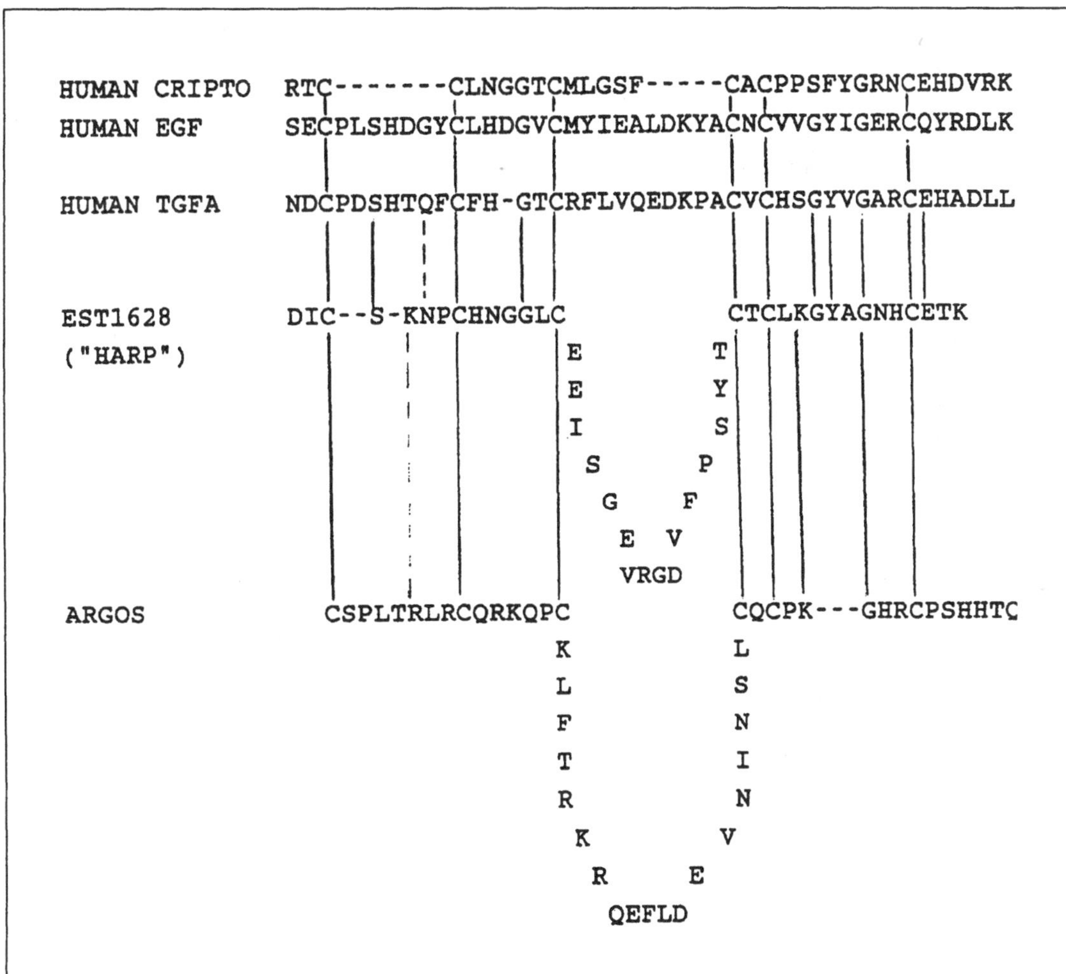

Fig. 5.11. Human EGF-related cytokines and Drosophila *Argos.*

Acknowledgments

I am grateful to my colleagues, in particular Tony Burgess, Hong He, Anjali Tikoo, Dennis Wang and Masha Fridman, who were involved in our recent, as yet unpublished, work. This work was supported in part by grants from the National Health and Medical Research Council (NHMRC) and the Anti-Cancer Council of Victoria (ACCV).

References

1. Trahey M, McCormick F. A cytoplasmic protein stimulates normal Ras GTPase, but does not affect oncogenic mutants. Science 1987; 238: 542-545.
2. Gibbs JB, Schaber MD, Allard WJ, Sigal IS, Scolnick EM. Purification of Ras GAP from bovine brain. Proc Natl Acad Sci USA 1988; 85:5026-5030.
3. Vogel US, Dixon RAF, Schaber MD, Diehl RE, Marshall MS, Scolnick EM, Sigal IS, Gibbs JB. Cloning of bovine GAP and its interaction with oncogenic Ras. Nature 1988; 335:90-93.
4. Marshall MS, Hill WS, Ng AS, Vogel US, Schaber MD, Scolnick EM, Dixon RAF, Sigal IS, Gibbs JB. A C-terminal domain of GAP is sufficient to stimulate Ras GTPase activity. EMBO J 1989; 8:1105-1110.
5. Maruta H, Holden J, Sizeland A, D'Abaco G. The residues of Ras and Rap proteins that determine their GAP specificities. J Biol Chem 1991; 266:11661-11668.
6. Briggs SD, Bryan SS, Jove R, Sanderson SD, Smithgall TE. The Ras GAP is an SH3 domain-binding protein and substrate for the Src-related Tyr kinase, Hck. J Biol Chem 1995; 270:14718-14724.
7. Parker F, Maurier F, Delumeau I et al. A Ras GAP SH3 binding protein (p68/G3BP). Mol Cell Biol 1996; 16:2561-2569.
8. Ellis C, Moran M, Anderson D, Liu X, Mbamalu G, Pawson T. Phosphorylation of GAP and GAP-associated proteins by transforming and mitoigenic tyrosine kinases. Nature 1990; 343:377-381.
9. Pawson T. Protein modules and signaling networks. Nature 1995; 373:573-580.
10. Garrett MD, Self AJ, van Oers C, Hall A. Identification of distinct cytoplasmic targets for Ras and Rho regulatory proteins. J Biol Chem 1989; 264:10-13.
11. Lancaster CA, Taylor-Harris PM, Self AJ, Brill S, van Erp HE, Hall A. Characterization of Rho GAP. A GAP for Rho-related small GTPases. J Biol Chem 1994; 269:1137-1142.
12. Lamarche N, Hall A. GAPs for Rho-related GTPases. TIG 1994; 10:436-440.
13. Settleman J, Albright CF, Foster LC, Weinberg RA. Association between GTPase activators for Rho and Ras families. Nature 1992; 359:153-154.
14. Kikuchi A, Sasaki T, Araki S, Hata Y, Takai Y. Purification and characterization from bovine brain cytosol of two GAPs specific for rap1A G

protein having the same effector domain as Ras. J Biol Chem 1989; 264:9133-9136.
15. Rubinfeld B, Munemitsu S, Clark R, Conroy L, Watt K, Crosier Wj, McCormick F, Polakis P. Molecular cloning of a GAP specific for rap1. Cell 1991; 69:1033-1042.
16. Xu G, O'Connell P, Viskochil D, Cawthorn R, Robertson M, Culver M, Dunn D, Stevens J, Gesteland R, White R, Wiess R. The neurofibromatosis type 1 gene encodes a protein related to GAP. Cell 1992; 62:599-608.
17. Martin G, Viskochil D, Bollag G, McCabe PC, Crosier W, Haubruck H, Conroy L, Clark R, O'Connell P, Cawthorn R, Innis MA, McCormick F. The GAP-related domain of the neurofibromatosis type 1 gene product interacts with Ras. Cell 1990; 63:843-849.
18. Xu G, Lin B, Tanaka K, Dunn D, Wood D, Gesteland R, White R, Wiess R, Tamanoi F. The catalytic domain of the neurofibromatosis type gene product stimulates Ras GTPase and complements Ira mutants of S. cerevisiae. Cell 1990; 63:835-841.
19. Nur-E-Kamal MSA, Varga M, Maruta H. The GTPase-activating NF1 fragment of 91 amino acids reverses v-Ha-Ras-induced malignant phenotype. J Biol Chem 1993; 268:22331-22337.
20. Fridman M, Tikoo A, Varga M, Murphy A, Nur-E-Kamal MSA, Maruta H. The Minimal fragments of c-Raf-1 and NF1 that can suppress v-Ha-Ras-induced malignant phenotype. J Biol Chem 1994; 269:30105-30108.
21. DeClue JE, Papageorge AG, Fletcher JA, Diehl SR, Ratner N, Vass WC, Lowy DR. Abnormal regulation of mammalian Ras contributes to malignant tumor growth in type 1 neurofibromatosis. Cell 1992; 69:265-273.
22. Basu TN, Gutman DH, Fletcher JA, Glover TW, Collins FS, Downward J. Aberrant regulation of Ras proteins in malignant tumor cells from type 1 neurofibromatosis patients. Nature 1992; 356:713-715.
23. Frech M, John J, Pizon V, Chardin P, Tavitian A, Clark R, McCormick F, Wittinghofer A. Inhibition of GAP stimulation of Ras GTPase by Rap1. Science 1990; 249:169-171.
24. Hata Y, Kikuchi A, Sasaki T, Schaber MD, Gibbs JB, Takai Y. Inhibition of GAP-stimulated Ras GTPase activity by rap1. J Biol Chem 1990; 265:7104-7107.
25. Holden J, Nur-E-Kamal MSA, Fabri L, Nice E, Hammacher A, Maruta H. Rsr1 and rap1 GTPases are activated by the same GAP and require Thr65 for their activation. J Biol Chem 1991; 266:16992-16995.
26. Nur-E-Kamal MSA, Maruta H. The role of Gln61 and Glu63 of Ras GTPases in their activation by NF1 and Ras GAP. Mol Biol Cell 1992; 3:1437-1442.
27. Nur-E-Kamal MSA, Sizeland A, D'Abaco G, Maruta H. Asp26, Glu31, Val45 and Tyr64 of Ras proteins are required for their oncogenicity. J Biol Chem 1992; 267:1415-1418.
28. Morii N, Kumagai N, Nur-E-Kamal MSA, Narumiya S, Maruta H. Rho GAP of 28 kDa (GAP2), but not of 190 kDa (p190), requires Asp65 and Asp67 of Rho GTPase for its activation. J Biol Chem 1993; 268:

27160-27163.
29. Maekawa M, Nakamura S, Hattori S. Purification of a novel GAP from rat brain. J Biol Chem 1993; 268:22948-22952.
30. Maekawa M, Li S, Iwamatsu A, Morishita T, Yokota K, Imai Y, Kohsaka S, Nakamura S, Hattori S. A novel Ras GAP which has phospholipid-binding and BTK homology regions. Mol Cell Biol 1994; 14:6879-6885.
31. Cullen PJ, Hsuan JJ, Truong O, Letcher AJ, Jackson TR, Dawson AP, Irving RF. Identification of a specific Ins (1,3,4,5)P_4-binding protein as a member of the GAP1 family. Nature 1995; 376: 527-530.
32. Hu DC, Kariya K, Tamada M, Akasaka K, Shirouzu M, Yokoyama S, Kataoka T. Cys-rich region of Raf-1 interacts with activator domain of post-translationally modified Ha-Ras. J Biol Chem 1995; 270:30274-30277.
33. Warne PH, Viciana PR, Downward J. Direct interaction of Ras and the N-terminal region of Raf-1 in vitro. Nature 1993; 364:352-355.
34. Maruta H, Burgess AW. RAS*—a common cancerous gene. Today's Life Science 1992; 4 (8): 24-32.
35. Bos JL. Ras oncogenes in human cancer: A review. Cancer Res 1989; 49:4682-4689.
36. Barbacid M. Ras genes. Annu Rev Biochem 1987; 56:779-827.
37. Lowy DR, Willumsen BM. Function and regulation of Ras. Annu Rev Biochem 1993; 62:851-891.
38. Gibbs JB, Schaber MD, Schofield TL, Scolnick EM, Sigal IS. *Xenopus* oocyte germinal-vesicle breakdown induced by Val12-Ras is inhibited by a cytosol-localized Ras mutant. Proc Natl Acad Sci USA 1989; 86:6630-6634.
39. Sigal IS, Gibbs JB, D'Alonzo JS, Scolnick EM. Identification of effector residues and a neutralizing epitope of Ha-Ras. Proc Natl Acad Sci USA 1986; 83:4725-4729.
40. Kitayama H, Sugimoto Y, Matsuzaki T, Ikawa Y, Noda M. A Ras-related gene with transformation suppressor activity. Cell 1989; 56:77-84.
41. Pizon V, Chardin P, Lerosey I, Olofsson B, Tavitian A. Human cDNA Rap1 and Rap2 homologous to the *Drosophila* gene DRAS3 encode proteins closely related to Ras in the effector region. Oncogene 1988; 3:201-204.
42. Kitayama H, Matsuzaka T, Ikawa Y, Noda M. Genetic analysis of Rap1: potentiation of its tumor suppressor activity by specific point mutations. Proc Natl Acad Sci USA 1990; 87:4284-4288.
43. Zhang K, Noda M, Vass WC, Papageorge AG, Lowr DR. Identification of small clusters of divergent amino acids that mediate the opposing effects of Ras and Rap1. Science 1990; 249:162-165.
44. Prendergast GC, Gibbs JB. Ras regulatory interactions: Novel targets for anti-cancer intervention? BioEssays 1994; 16:187-191.
45. Maruta H, Burgess AW. Regulation of the Ras signaling network. BioEssays 1994; 16:489-496.
46. Zhang XF, Settleman J, Kyriakis JM, Takeuchi-Suzuki E, Elledge SJ,

Marshall MS, Bruder JT, Rapp UR, Avruch J. Normal and oncogenic Ras proteins bind to the N-terminal regulatory domain of c-Raf-1. Nature 1993; 364:308-313.
47. Vojtek AB, Hollenberg SM, Cooper JA. Mammalian Ras interacts directly with the Ser/Thr kinase Raf. Cell 1993; 74:205-214.
48. Yamamori B, Kuroda S, Shimizu K, Fukui K, Ohtsuka T, Takai Y. Purification of a REKS from bovine brain cytosol and its identification as a complex of B-Raf and 14.3.3. J Biol Chem 1995; 270:11723-11726.
49. Mark GF, MacIntyre RJ, Digan ME, Ambrosio L, Perrimon N. *Drosophila* homologue of the Raf oncogene. Mol Cell Biol 1987; 7:2134-2140.
50. Jansen HW, Lurz R, Bister K, Bonner TI, Mark GE, Rapp UR. Homologous cell-derived oncogenes in avian carcinoma virus MH2 and murine sarcoma virus 3611. Nature 1984; 307:281-284.
51. Rapp UR. Role of Raf-1 Ser/Thr kinase in growth factor signal transduction. Oncogene 1991; 6:495-500.
52. Daum G, Eisenmann-Tappe I, Fries HW, Troppmair J, Rapp UR. The ins and outs of Raf kinases. Trends in Biochem Soc 1994; 19:474-480.
53. Kyriakis JM, App H, Zhang XF, Banerjee P, Brautigan DL, Rapp UR, Avruch J. Raf-1 activates MAP kinase-kinase. Nature 1992; 358:417-421.
54. McCormick F. Raf: the Holy Grail of Ras biology? Trends in Cell Biol 1994; 4:3347-350.
55. Cowley S, Paterson H, Kemp P, Marshall CJ. Activation of MEK is necessary and sufficient for PC12 differentiation and for transformation of NIH/3T3 cells. Cell 1994; 77:841-852.
56. Stokoe D, MacDonald SG, Cadwallader K, Symons M, Hancock JF. Activation of Raf as a result of recruitment to the plasma membranes. Science 1994; 264:1463-1467.
57. Leevers SJ, Paterson HF, Marshall CJ. Requirement for Ras in Raf activation is overcome by targeting Raf to the plasma membranes. Nature 1994; 369; 411-414.
58. Qiu RG, Che J, Kirn D, McCormick F, Symons M. An essential role for Rac in Ras transformation. Nature 1995; 374:457-459.
59. Beranger F, Goud B, Tavitian A, de Gunzburg J. Association of rap1 with the Golgi complex. Proc Natl Acad Sci USA 1991; 88:1601-1610.
60. Therrien M, Chang HC, Solomom NM, Karim FD, Wassarman DA, Rubin GM. KSR, a novel protein kinase required for RAS signal transduction. Cell 1995; 83:879-888.
61. Stanton VP, Nichols DW, Laudano AP, Cooper GM. Definition of the human Raf N-terminal regulatory region by deletion mutagenesis. Mol Cell Biol 1989; 9:639-647.
62. Fabian JR, Daar IO, Morrison DK. Critical Tyr residues regulate the enzymatic and biological activity of Raf-1 kinase. Mol Cell Biol 1993; 13:7170-7179.
63. Fabian JR, Vojtek AB, Cooper JA, Morrison DK. A single amino acid change in Raf-1 inhibits Ras binding and alters Raf-1 function. Proc Natl Acad Sci USA 1994; 91:5982-5986.

64. Melnick MB, Perkins LA, Lee M, Ambrosio L, Perrimon N. Developmental and molecular characterization of mutations in the *Drosophila* Raf Ser/Thr kinase. Development 1993; 118:127-138.
65. Barnard D, Diaz B, Hettich L, Chuang E, Zhang ZF, Avruch J, Marshall M. Identification of the sites of interaction between Raf and Ras-GTP. Oncogene 1995; 10:1283-1290.
66. Nassar N, Horn G, Herrmann C, Scherer A, McCormick F, Wittinghofer A. The 2.2 A crystal structure of the RAS binding domain of the Ser/Thr kinase Raf in complex with rap1A and a GTP analogue. Nature 1995; 375:554-560.
67. Shirouzu M, Koide H, Fujita-Yoshigaki J, Oshio H, Toyama Y, Yamazaki K, Fuhrman SA, Villafranca E, Kaziro Y, Yokoyama S. Mutations that abolish the ability of Ha-Ras to associate with Raf. Oncogene 1994; 9:2153-2157.
68. Hiles ID, Otsu M, Volinia S, Fry MJ, Gout I, Dhand R, Panayotou G, Ruiz-Larrea F, Thompson A, Totty NF, Hsuan JJ, Courtneidge SA, Parker PJ, Waterfield MD. Phosphatidylinositol 3-kinase: structure and expression of the p110 catalytic subunit. Cell 1992; 70:419-429.
69. Rodriguez-Viciana P, Warne PH, Dhand R, Vanhaesebroeck B, Gout I, Fry MJ, Waterfield MD, Downward J. Pi-3 kinase as a direct target of Ras. Nature 1994; 370:527-532.
70. Zhang QX, Davis ID, Baldwin GS. Controlled overexpression of selected domains of the p85 subunit of PI-3 kinase reverses v-Ha-Ras transformation. Biochim Biophys Acta 1996; (in press).
71. Martin GA, Yatani A, Clark R, Conroy L, Polakis P, Brown AM, McCormick F. GAP domains responsible for Ras-dependent inhibition of muscarinic atrial K^+ channel currents. Science 1992; 255:192-194.
72. Henkemeyer M, Rossi DJ, Holmyard DP, Puri MC, Mbamalu G, Harpal K, Shih TS, Jacks T, Pawson T. Vascular system defects and neuronal apoptosis in mice lacking Ras GAP. Nature 1995; 377:695-701.
73. McGlade J, Brunkhorst B, Anderson D, Settleman J, Dedhar S, Rozakis-Adock M, Chen LB, Pawson T. The N-terminal region of GAP regulates cytoskeletal structure and cell adhesion. EMBO J 1992; 12:3073-3081.
74. Chang JS, Kobayashi M, Wang DZM, Maruta H, Iwashita S. Two regions with differential growth-modulating activity in the N-terminal domain of Ras GAP: Src homology and Gly-Ala-Pro-rich regions. Eur J Biochem 1995; 232:691-699.
75. Duchesne M, Schweighoffer F, Parker F, Clerc F, Frobert Y, Thang MN, Tocque B. Identification of the SH3 domain of GAP as an essential sequence for Ras-GAP-mediated signaling. Science 1993; 259:525-528.
76. Zhao JF, Nakano H, Sharma S. Suppression of RAS and MOS transformation by radicicol. Oncogene 1995; 11:161-173.
77. Pronk GJ, de Vries-Smits AMM, Ellis C, Bos JL. Complex formation between Ras GAP and phosphoproteins p62 and p190 is independent of Ras signaling. Oncogene 1993; 8:2773-2780.
78. Foster R, Hu KQ, Shaywitz DA, Settleman J. p190 Rho GAP, the major

Ras GAP-associated protein, binds GTP directly. Mol Cell Biol 1994; 14:7173-7181.

79. Lebowitz PF, Davide JP, Prendergast GC. Evidence that farnesyl-transferase inhibitors suppress Ras transformation by interfering with Rho activity. Mol Cell Biol 1995; 15:6613-6622.
80. Coso OA, Chiariello M, Yu JC, Teramoto H, Crespo P, Xu N, Miki T, Gutkind JS. The small G proteins Racl and CDC42 regulate the activity of the JNK/SAPK signaling pathway. Cell 1995; 81:1137-1146.
81. Minden A, Lin A, Claret FX, Abo A, Karin M. Selective activation of the JNK signaling cascade and c-Jun transcriptional activity by the small GTPases Rac and CEC42. Cell 1995; 81:1147-1157.

81a. Hill CS, Wynne J, Treisman R. The Rho family GTPases RhoA, Racl and CDC42 regulate transcriptinal activation by SRF. Cell 1995; 81:1159-1170.

82. Ono Y, Fujii T, Ogita K, Kikkawa U, Igarashi K, Nishizuka Y. Protein kinase C ζ-subspecies from rat brain: its structure, expression and properties. Proc Natl Acad Sci USA 1989; 86:3099-3103.
83. Dominguez I, Diaz-Meco MT, Municio MM, Berra E, de Herreros GA, Cornet ME, Sanz L, Moscat J. Evidence for a role of protein kinase C ζ-subspecies in maturation of *Xenopus* oocytes. Mol Cell Biol 1992; 12:3776-3783.
84. Diaz-Meco MT, Lozano J, Municio MM, Berra E, Frutos S, Sanz L, Moscat J. Evidence for the in vitro and in vivo interaction off Ras with protein kinase C-ζ. J Biol Chem 1994; 269:31706-31710.
85. Bjoerkoey G, Oevervatn A, Diaz-Meco MT, Moscat J, Johansen T. Evidence for a bifurcation of the mitogenic signaling pathway activated by Ras and PC-PLase C. J Biol Chem 1995; 270: 21299-21306.
86. Chardin P, Tavitian A. Coding sequences of human RalA and RalB cDNAs. Nucleic Acids Res 1989; 17:4380.
87. Olofsson B, Chardin P, Touchot N, Zahraoui A, Tavitian A. Expression of the Ras-related RalA, Rho12 and Rab genes in adult mouse tissues. Oncogene 1988; 3:231-234.
88. Albright CF, Giddings BW, Liu J. Vito M, Weinberg RA. Characterization of a GDP-dissociation stimulator for a Ras-related GTPase (Ral). EMBO J 1993; 12:339-347.
89. Kikuchi A, Demo SD, Ye ZH, Chen YW, Williams LT. RalGDS family members interact with the effector loop of Ras. Mol Cell Biol 1994; 14:7483-7491.
90. Hofer F, Fields S, Schneider C, Martin GS. Activated Ras interacts with the Ral GDS. Proc Natl Acad Sci USA 1994; 91: 11089-11093.
91. Urano T, Emkey R, Feig LA. Ral GTPases mediate a distinct downstream signaling pathway from Ras that facilitates cellular transformation. EMBO J 1996; 15:810-816.
92. Jiang H, Luo JQ, Urano T, Frankel P, Lu Z, Foster DA, Feig LA. Involvement of Ral GTPase in v-Src-induced phospholipase D activation. Nature 1995; 378:409-412.

93. Carnero A, Cuadrado A, de Peso L, Lacal JC. Activation of type D phospholipase by serum stimulation and Ras-induced transformation in NIH/3T3 cells. Oncogene 1994; 9:1387-1395.
94. Cantor SB, Urano T, Feig LA. Identification and characterization of Ral-binding protein 1, a potential downstream target of Ral GTPases. Mol Cell Biol 1995; 15:4578-4584.
95. Jullien-Flores V, Dorseuil O, Romero F, Letourneur F, Saragosti S, Berger R, Tavitian A, Gacon G, Camonis JH. Bridging Ral GTPase to Rho Pathways. RLIP76, a Ral effector with CDC42/Rac GAP activity. J Biol Chem 1995; 270:22473-22477.
96. Park SH, Weinberg RA. A putative effector of Ral has homology to Rho/Rac GAPs. Oncogene 1995; 11:2349-2355.
97. Weber K, Lazarides E, Goldman RD, Vogel A, Pollack R. Localization and distribution of actin fibers in normal, transformed and revertant cells. Cold Spring Harbor Symp Quant Biol 1974; 39:363-369.
98. Pollard TD, Cooper JA. Actin and actin-binding proteins. A critical evaluation of mechanisms and functions. Annu Rev Biochem 1986; 55:987-1035.
99. Fernandez JLR, Geiger B, Salomon D, Sabanay I, Zoeller M, Ben-Ze'ev A. Suppression of tumorigenicity in SV40-transformed cells after transfection with vinculin cDNA. J Cell Biol 1992; 119: 427-438.
100. Glueck U, Kwiatkowski D, Ben-Ze'ev A. Supression of tumorigenicity in SV40-transformed 3T3 cells transfected with α-actinin cDNA. Proc Natl Acad Sci USA 1993; 90:383-387.
101. Vandekerckhove J, Bauw G, Vancompernolle A, Honore B, Celis J. Comparative two-dimensional gel analysis and microsequencing identifies gelsolin as one of the most prominent down-regulated markers of transformed human fibroblast and epithelial cells. J Cell Biol 1990; 111:95-102.
102. Muellauer L, Fujita H, Shizaki A, Kuzumaki N. Tumor-suppressive function of mutated gelsolin in Ras-transformed cells. Oncogene 1993; 8:2531-2536.
103. Yin HL. Gelsolin: calcium and polyphosphoinositide-regulated actin modulating protein. BioEssays 1987; 7:176-179.
104. Tikoo A, Varga M, Ramesh V, Guesella J, Maruta H. An anti-Ras function of neurofibromatosis type 2 gene product (NF2/Merlin). J Biol Chem 1994; 269:23387-23390.
105. Tikoo A, Lo SH, Chen LB, Moriyama K, Yahara I, Maruta H. Tensin, cofilin mutants and cytochalasins block an oncogenic Ras signaling by capping the barbed end of actin filament. Cancer Res 1996; (submitted).
106. He H, Watanabe T, Maruta H. Human HS1 (hemotopoiesis specific) requires both its SH3 domain and F-actin binding domain for a tumor suppressor activity. Cell 1996; (submitted).
107. Lo SH, Weisberg E, Chen LB. Tensin: a potential link between the cytoskeleton and signal transduction. BioEssays 1994; 16:817-823.
108. Maruta H, He H. Cytoskeletal SH3 proteins. Biochemistry (Tokyo) 1995; 67:1210-1217.

109. Fankhauser C, Reymond A, Cerutti L, Utzig S, Hofmann K, Simanis V. The S. pombe CDC15 gene is a key element in the regulation of F-actin at mitosis. Cell 1995; 82:435-444.
110. Taniuchi I, Kitamura D, Maekawa Y, Fukuda T, Kishi H, Watanabe T. Antigen-receptor induced clonal expansion and deletion of lympho-cytes are impaired in mice lacking HS1, a substrate of the antigen-receptor-coupled Tyr kinases. EMBO J 1995; 14; 3664-3678.
111. Kariya K, Kataoka T. RAS proteins in S. cerevisiae. Exp Med (Tokyo) 1996; 14:259-260.
112. Orloff GJ, Allen PG, Miklos GLG, Young IG, Campbell HD, Kwiatkowski DJ. Human flightless-I has actin binding ability. 1996; (submitted).
113. Minato T, Wang J, Akasaka K, Okada T, Suzuki N, Kataoka T. Quantitative analysis of mutually competitive binding of human Raf and yeast adenylate cyclase to Ras. J Biol Chem 1994; 269:20845-20851.
114. Campbell HD, Schimanski T, Claudianos C, Ozsarac N, Kasprzak AB, Costell JN, Young IG, De Couet HG, Miklos GLG. The *Drosophila* flightless-I gene involved in gastrulation and muscle degeneration encodes gelsolin-like and Leu-rich repeat domains and is conserved in *C. elegans* and humans. Proc Natl Acad Sci USA 1993; 90:11386-11390.
115. Chen KS, Gunaratne PH, Hoheisel JD, Young IG, Miklos GLG, Greenberg F, Shaffer LG, Campbell HD, Lupski JR. The human homologue of the *drosophila* flightless-I gene maps within the Smith-Magenis microdeletion critical region in 17p11.2. Am J Hum Genet 1995; 56:175-182.
116. Ridley AJ, Paterson HF, Johnson CL, Diekman D, Hall A. Rac regulates growth factor-induced membrane ruffling. Cell 1992; 70:401-410.
117. Downward J, de Gunzburg J, Riehl R, Weinberg RA. Ras-induced responsiveness of phosphatidylinositol turnover to bradykinin is a receptor number effect. Proc Natl Acad Sci USA 1988; 85: 5774-5778.
118. Kozma R, Ahmed S, Best A, Lim L. CDC42 and bradykinin promote formation of peripheral microspikes and filopodia in Swiss 3T3 fibroblasts. Mol Cell Biol 1995; 15:1942-1952.
119. Ridley AJ. Membrane ruffling and signal transduction. BioEssays 1994; 16:321-327.
120. Khosravi-Far R, Solski PA, Clark GJ, Kinch MS, Der CJ. Activation of Rac1, RhoA and MAP kinase is required for Ras transformation. Mol Cell Biol 1995; 15:6443-6453.
121. Ridley AJ, Hall A. Rho regulates the assembly of focal adhesions and actin stress fibers in response to growth factors. Cell 1992; 70:389-399.
122. Reinhard J, Scheel AA, Diekmann D, Hall A, Ruppert C, Baehler M. A novel type of myosin implicated in signaling by Rho family GTPases. EMBO J 1995; 14:697-704.
123. Pollard TD, Doberstein SK, Zot HG. Myosin I. Annu Rev Physiol 1991; 51:653-681.
124. Hartwig JH, Bokoch GM, Carpenter CL, Janmey PA, Taylor LA, Toker A, Stossel TP. Thrombin receptor ligation and activated Rac uncap actin

filament barbed ends through phosphoinositide synthesis in permeabilized human platelets. Cell 1995; 82:643-653.
125. Moriyama K, Yonezawa N, Sakai H, Yahara I, Nishida E. Mutational analysis of an actin-binding site of cofilin and characterization of chimeric cofilin/destrin proteins. J Biol Chem 1992; 267: 7240-7244.
126. Yahara I, Harada F, Sekita S, Yoshihira K, and Natori S. Correlation between effects of 24 different cytochalasins on cellular structures and events and those on actin in vitro. J Cell Biol 1982; 92:69-78.
127. Buettner R, Yim SO, Hong YS, Boncinelli E, Tainsky M. Alteration of homeobox gene expression by N-Ras transformation of PA-1 human teratocarcinoma cells. Mol Cell Biol 1991; 11:3573-3583.
128. Yoshida M, Horinouchi S, Beppu T. Trichostatin A and trapoxin: novel chemical probes for the role of histone acetylation in chromatin structure and function. BioEssays 1995; 17:423-430.
129. Sugita K, Koizumi K, Yoshida H. Morphological reversion of Sis-transformed NIH/3T3 cells by trichostatin A. Cancer Res 1992; 52:168-172.
130. Futamura M, Monden Y, Okabe T, Fujita-Yoshigaki J., Yokoyama S, Nishimura S. Trichostatin A inhibits both Ras-induced neurite outgrowth of PC12 cells and morphological transformation of NIH/3T3 cells. Oncogene 1995; 10:1119-1123.
131. Higashiyama S, Abraham JA, Miller J, Fiddes JC, Klagsbrun M. A heparin-binding growth factor secreted by macrophage-like cells is related to EGF. Science 1991; 251:936-939.
132. Schweitzer R, Howes R, Smith R, Shilo BZ, Freeman M. Inhibition of *Drosophila* EGF receptor activation by the secreted protein Argos. Nature 1995; 376:699-702.
133. Sekine A, Fujiwara M, Narumiya S. Asp residue in the Rho is the modification site for Botulinum ADP-ribosyltransferase C3. J Biol Chem 1989; 264:8602-8605.
134. Groenen LC, Nice EC, Burgess AW. Structure-function relationships for the EFG/TGF-α family of mitogens. Growth Factors 1994; 11:235-257.
135. McCarthy SA, Samuels ML, Pritchard CA, Abraham JA, McMahon M. Rapid induction of HB-EGF/diphtheria toxin receptor expression by Raf and RAS oncogenes. Genes & Dev 1995; 9:1953-1964.
136. Grugel S, Finkenzeller G, Weindel K, Barleon B, Marme D. Both v-Ha-Ras and v-Raf stimulate expression of the VEGF in NIH/3T3 cells. J Biol Chem 1995; 270:25915-21919.
137. Dlugosz AA, Cheng C, Williams EK et al. Autocrine TGF-α is dispensable for v-Ha-Ras-induced epidermal neoplasia: potential involvement of alternative EGF receptor ligands. Cancer Res 1995; 55:1883-1893.
138. Naglich JG, Metherall JE, Russell DW, Eidels L. Expression cloning of a diphtheria toxin receptor: identity with an HB-EGF precurson. Cell 1992; 69:1051-1061.
139. Mitamura T, Higashiyama S, Taniguchi N, Klagsbrun M, Mekada E. Diphtheria toxin (DT) binds to the EGF-like domain of human HB-EGF/DT receptor and inhibits specifically its mitogenic activity. J Biol

Chem 1995; 270:1015-1019.
140. Goishi K, Higashiyama S, Klagsbrun M, Nakano N, Umata T, Ishikawa M, Mekada E, Taniguchi N. Phorbol ester induces the rapid processing of celll surface HB-EGF: conversion from juxtacrine to paracrine growth factor activity. Mol Biol Cell 1995; 6:967-980.
141. Buzzi S, Maistrello I. Inhibition of growth of Ehrlich tumors in Swiss Mice by diphtheria toxin. Cancer Res 1973; 33:2349-2353.
142. Mayer U, Nuesslein-Volhard C. A group of genes required for pattern formation in the ventral ectoderm of the *Drosophila* embryos. Genes & Dev 1988; 2:1496-1511.
143. Rutledge B, Zhang K, Bier E, Jan YN, Perrimon N. The *Drosophila* spitz gene encodes a putative EGF-like growth factor involved in dorsal-ventral axis formation and neurogenesis. Genes & Dev 1992; 6:1503-1517.
144. Freeman M, Klaembt C, Goodman CS, Rubin GM. The Argos gene encodes a diffusible factor that regulates cell fate decision in the *Drosophila* eye. Cell 1992; 69:963-975.
145. Kretzschmar D, Brunner A, Wiersdorff V, Pflugfelder GO, Heisenberg M, Schneuwly S. Giant lens, a gene involved in cell determination and axon guidance in the visual system of *Drosophila*. EMBO J 1992; 11:2531-2539.
146. Yamamoto T, Matsui T, Nakafuku M, Iwamatsu A, Kaibuchi K. A novel GAP for R-Ras. J Biol Chem 1995; 270:30557-30561.
147. Kumar CC, Chang C. Human smooth muscle myosin light chain-2 gene expression is repressed in Ras transformed fibroblast cells. Cell Growth & Differ 192; 3:1-10.
148. Konieczny SF, Drobes BL, Menke SL, Taparowsky EJ. Inhibition of myogenic differentiation by the H-Ras ioncogene is associated with the down regulation of the MyoD1 gene. Oncogene 1989; 4:473-481.
149. Lassar AB, Thayer MJ, Overell RW, Weintraub H. Transformation by activated Ras or Fos prevents myogenesis by inhibiting expression of MyoD1. Cell 1989; 58:659-667.
150. Avvedimento VE, Musti AM, Ueffing M. et al. Reversible inhibition of a thyroid-specific trans-acting factor by Ras. Gene & Develop 1991; 5:22-28.
151. Krzyzosiak WJ, Shindo-Okada N, Teshima H, Nakajima K, Nishimura S. Isolation of genes specifically expressed in flat revertant cells derived from Ras-transformed NIH 3T3 cells by treatment with azatyrosine. Proc Natl Acad Sci USA 1992; 89: 4879-4883.
152. Contente S, Kenyon K, Rimoldi D, Friedman RM. Expression of gene RRG is associated with reversion of NIH 3T3 cells by LTR-c-Ha-Ras. Science 1990; 249:796-798.
153. Sugimoto Y, Ikawa, Y, Nakauchi H. Thy-1 as a negative regulator in Ras-transformed mouse fibroblasts. Cancer Res 1991; 51:99-104.
154. Lin X, Nelson PJ, Frankfort B et al. Isolation and characterization of a novel mitogenic regulatory gene, 322, which is transcriptionally suppressed in cells transformed by Src and Ras. Mol Cell Biol 1995; 15:2754-2762.
155. Glick AB, Sporn MB, Yuspa SH. Altered regulation of TGF-β1 and

TGF-α in primary keratinocytes and papillomas expressing v-Ha-Ras. Mol Carcinog 1991; 4:210-219.
156. Normanno N, Selvam MP, Qi CF et al. Amphiregulin as an autocrine growth factor for c-Ha-Ras/c-ErbB-2-transformed human mammary epithelial cells. Proc Natl Acad Sci USA 1994; 91:2790-2794.
157. Liang P, Averboukh L, Zhu W, Pardee AB. Ras activation of genes: Mob-1 as a model. Proc Natl Acad Sci USA 1994; 91:12515-12519.
158. Winston JT, Coats SR, Wang YZ, Pledger WJ. Regulation of cell cycle machinery by oncogenic RAS. Oncogene 1996; 12:127-134.
159. Medema RH, Wubbolts R, Bos JL. Two dominant inhibitory mutants of Ras interfere with insulin-induced gene expression. Mol Cell Biol 1991; 11:5963-5967.
160. Cornwell MM, Smith DE. A signal transduction pathway for activation of the MDR1 promoter involves the c-Raf-1. J Biol Chem 1993; 268:15347-15350.
161. Lin CQ, Dempsey PJ, Coffey RJ, Bissell MJ. Extracellular matrix regulates whey acidic protein gene expression by suppression of TGF-α in mouse mammary epithelial cells: studies in culture and in transgenic mice. J Cell Biol 1995; 129:1115-1126.
162. Hajnal A, Klemenz R, Schaefer R. Substraction cloning of H-Rev107, a gene specifically expressed in H-Ras resistant fibroblasts. Oncogene 1994; 9:479-490.
163. Kumar CC, Prorock-Rogers C, Kelly J, Dong Z, Lin JJ, Armstrong L, Kung HF, Weber MJ, Afonso A. SCH51344 inhibits Ras transformation by a novel mechanism. Cancer Res 1995; 55:5106-5117.

CHAPTER 6

From RAS to MAPK: Cell-Free Assay System for RAS- and Rap1-Dependent B-Raf Activation

Kazuya Shimizu, Toshihisa Ohtsuka and Yoshimi Takai

I. INTRODUCTION

It is well established that RAS activates the MAP kinase cascade, consisting of the mitogen-activated protein (MAP) kinase/extracellular signal-regulated kinase (ERK), ERK kinase (MEK) and MEK kinase in mammals (for reviews, see refs. 1-4). MAP kinase is phosphorylated at both serine/threonine and tyrosine residues by MEK, and this phosphorylation causes MAP kinase activation.[1-3] MEK is also phosphorylated at serine/threonine residues by MEK kinase and this phosphorylation causes MEK activation.[5-7] Many MEK kinases have been identified: these include c-Raf-1,[8-10] B-Raf,[11-15] Mos[16,17] and mSte11.[11,18] There are several lines of evidence that RAS is an upstream regulator of c-Raf-1: (i) anti-sense c-Raf-1 RNA and dominant negative c-Raf-1 inhibit the RAS-induced DNA synthesis and growth;[19] (ii) RAS genetically positions upstream of c-Raf-1 in *Drosophila*[20] and *Caenorhabditis elegans*;[21] (iii) dominant negative c-Raf-1 inhibits the RAS-induced MAP kinase activation in intact cells;[22] and (iv) GTP-RAS directly interacts with c-Raf-1.[23-28] Similarly, there are also several lines of evidence that RAS is an upstream regulator of B-Raf: (i) GTP-RAS directly interacts with B-Raf;[13,14] (ii) MAP kinase is activated by B-Raf, but not by c-Raf-1, in intact PC12 cells in response to NGF;[13] (iii) the association of MEK

Regulation of the RAS Signaling Network, edited by Hiroshi Maruta and Antony W. Burgess.

with immobilized RAS is dependent on B-Raf, but not on c-Raf-1, in rat brain;[14] and (iv) the MEK-activating activity is accompanied by B-Raf, but not c-Raf-1, in bovine brain or chromaffin cells.[15] However, it has not been shown that GTP-RAS directly activates these MEK kinases in a cell-free assay system using the purified samples of RAS and MEK kinases.

To identify the direct target molecule of RAS, we have developed a cell-free assay system in which GTP-RAS activates the MAP kinase cascade.[29] Using this assay system, we have found that GTP-RAS directly interacts with B-Raf and activates it in a cell-free system.[30] Moreover, we have found that GTP-Rap1, which shares 100% sequence identity in the effector domain with GTP-RAS,[31-33] also activates B-Raf in a cell-free system.[34]

II. CELL-FREE ASSAY SYSTEM FOR THE RAS-DEPENDENT ACTIVATION OF THE MAP KINASE CASCADE IN *XENOPUS* OOCYTE CYTOSOL

Immature *Xenopus laevis* oocytes (stage VI) were obtained surgically from hypothermically anesthetized females without collagenase treatment. The immature *Xenopus* oocytes were homogenized and centrifuged to prepare the cytosol and membrane fractions. Immunoblot analysis using the anti-RAS antibody revealed that the membrane fraction, but not the cytosol, contained RAS.[35] The *Xenopus* MAP kinase activity of the cytosol was assayed by the in-gel kinase assay method (Fig. 6.1A).[35] This assay is sensitive enough to detect the MAP kinase activity, even in a crude protein preparation.[36-38] The lipid-modified form of human recombinant Ki-RAS was purified from the membrane fraction of insect cells, which were infected with baculovirus carrying the cDNA of Ki-RAS.[39] GDP-Ki-RAS and guanosine 5'-(3-*O*-thio)triphosphate (GTPγS)-Ki-RAS were prepared[39] and their effect on the MAP kinase activity was examined in the cytosol fraction of *Xenopus* oocytes. GTPγS-Ki-RAS activated a protein kinase of about 41 kDa, whereas GDP-Ki-RAS was far less effective (Fig. 6.1B, lanes 1, 3, 5 and 7). This protein kinase was identified to be MAP kinase by immunoblot analysis. It may be noted that MAP kinase in the cytosol

Fig. 6.1 (opposite). Activation of MAP kinase and GST-ERK2 by Ki-RAS in the presence of the cytosol fraction of Xenopus *oocytes. The cytosol fraction of immature oocytes (200 μg of protein) was incubated at 30°C for 10 min with 300 nM of the lipid-modified form of GDP- or GTPγS-Ki-RAS in the presence or absence of GST-ERK2. The MAP kinase activity was detected by the in-gel kinase assay. (A) the procedure for a cell-free assay system. (B) the activation of MAP kinase and GST-ERK2. Lanes 1, 3, 5 and 7 in the absence of GST-ERK2; lanes 2, 4, 6 and 8, in the presence of GST-ERK2. Lanes 1 and 2, with GDP alone; lanes 3 and 4, with GTPγS alone; lanes 5 and 6, with lipid-modified GDP-Ki-RAS; lanes 7 and 8, with lipid-modified GTPγS-Ki-RAS. The white arrowhead, the black arrowhead and the black arrow indicate the positions of* Xenopus *MAP kinase, GST-ERK2 and S6 kinase II, respectively. The results shown are representative of three independent experiments.*

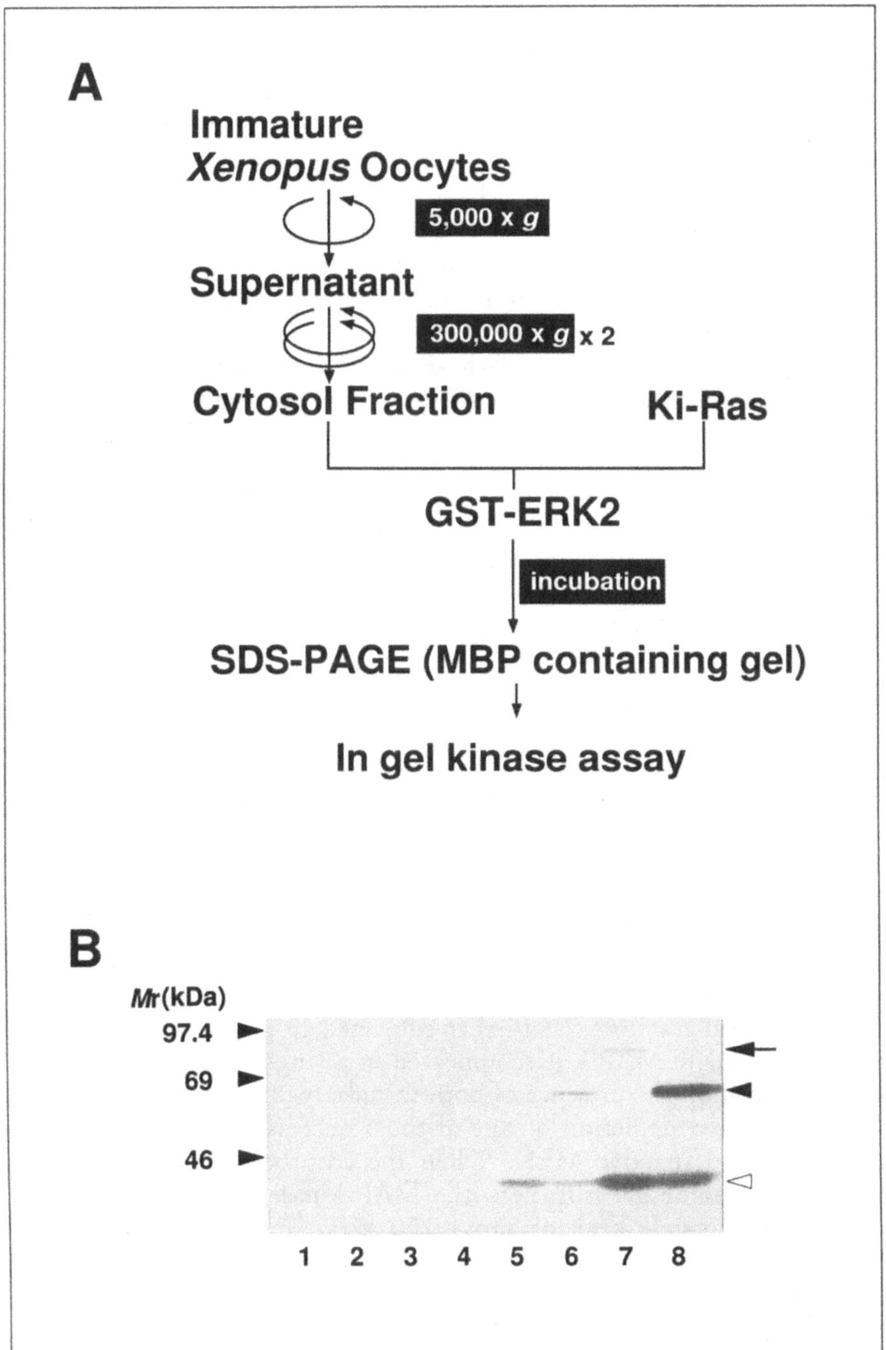
A
Immature
Xenopus Oocytes
5,000 x g
Supernatant
300,000 x g x 2
Cytosol Fraction
Ki-Ras
GST-ERK2
incubation
SDS-PAGE (MBP containing gel)
In gel kinase assay
B
Mr(kDa)
97.4
69
46
1 2 3 4 5 6 7 8

fraction showed basal kinase activity (Fig. 6.1B, lanes 1 and 3). This might be due to the partial activation of MAP kinase during the preparation of the cytosol fraction. Similar results were obtained when the lipid-modified form of human recombinant Ha-RAS was used. GTPγS-Ki-RAS activated a protein kinase of about 90 kDa in addition to MAP kinase (Fig. 6.1B, lane 7). This protein kinase may be S6 kinase II, which is known to be a possible substrate of MAP kinase and to be activated by its phosphorylation.[40]

To examine the effect of Ki-RAS on the exogenously added latent form of MAP kinase, the latent form of recombinant glutathione *S*-transferase (GST)-ERK2 fusion protein (GST-ERK2) of about 70 kDa was expressed in *E. coli*. GTPγS-Ki-RAS activated GST-ERK2 in addition to endogenous *Xenopus* MAP kinase in the presence of the cytosol fraction (Fig. 6.1B, lanes 2, 4, 6 and 8), but not in the absence of the cytosol. GDP-Ki-RAS was far less effective in this capacity (Fig. 6.1B, lane 6). It may be noted that GST-ERK2 showed weak kinase activity in the presence of the cytosol, but not in its absence. This might be due to the partial activation of MEK during the preparation of the cytosol. Similar results were obtained when the lipid-modified form of Ha-RAS was used.

III. REKS

Identification of REKS as a RAS-dependent MEK Kinase

The cytosol of immature oocytes was separated by DEAE-Sephacel column chromatography into unadsorbed and adsorbed fractions.[29] Immunoblot analysis by use of the anti-MAP kinase antibody revealed that the adsorbed fraction contained endogenous MAP kinase. GTPγS-Ki-RAS activated endogenous MAP kinase and GST-ERK2 only in the presence of both fractions, but not in the presence of either fraction alone. GDP-Ki-RAS was nearly inactive in this capacity. When the unadsorbed fraction was further subjected to gel filtration, the activity to stimulate MAP kinase appeared in a single peak. This activity was dependent on the presence of both the adsorbed fraction and GTPγS-Ki-RAS, and the depletion of one of them decreased the activity. This peak contained inactive MEK. When the adsorbed fraction was subjected to the same gel filtration, the MAP kinase stimulation activity appeared in a single peak of about 150 kDa. This activity was completely dependent on the presence of both MEK and GTPγS-Ki-RAS; the depletion of either of them decreased the activity to a near basal level. This activity was abolished by treatment with tryptic digestion or boiling, indicating that this activity is due to a protein molecule. We tentatively named this protein REKS (RAS-dependent ERK-Kinase Stimulator). The REKS peak did not contain MEK or MAP kinase.

Characterization of REKS

We have modified a cell-free assay system by use of GST-MEK and GST-ERK2 (Fig. 6.2A), partially purified REKS from the cytosol of the activated *Xenopus* eggs by Mono Q column chromatography and characterized it.[41] REKS forms a stable complex with GTPγS-RAS. REKS was coimmunoprecipitated with GTPγS-Ki-RAS and GTPγS-Ha-RAS, but not with GDP-Ki-RAS or GDP-Ha-RAS, by use of the anti-RAS antibody. REKS adsorbed to a GTPγS-GST-Ha-RAS-coupled glutathione agarose column, but not to a GDP-GST-Ha-RAS-coupled glutathione agarose column, and was coeluted with GTPγS-GST-Ha-RAS by reduced glutathione. The minimum Mr of REKS was estimated to be about 98 kDa on sodium dodecyl sulfate-polyacrylamide gel electrophoresis (SDS-PAGE). REKS phosphorylated this 98 kDa protein as well as GST-MEK. REKS was not recognized by any of the anti-c-Raf-1, anti-Mos and anti-mSte11 antibodies. These results indicate that REKS is a RAS-dependent MEK kinase different from these MEK kinases, but we have not yet identified it. The partially purified sample of REKS showed kinetic results essentially similar to those of the cytosol of *Xenopus* oocytes.

Synergistic Activation of REKS by 14-3-3 and RAS

The 14-3-3 family proteins were originally isolated by the systematic analysis of brain proteins as a series of acidic proteins of around 30 kDa and an isoelectric point of around 5 (for reviews, see refs. 42, 43). Subsequent studies have shown that the proteins of this family exist in dimeric form.[44,45] Seven isoforms have been identified; these members are expressed in most tissues and are particularly abundant in brain.[43] Multiple functions of 14-3-3 have been reported: an activator of tyrosine and tryptophan hydroxylases,[45] a potent inhibitor or activator of protein kinase C,[46,47] a stimulatory regulator (Exo1) for Ca^{2+}-dependent exocytosis in permeabilized adrenal chromaffin cells[48] and phospholipase A_2 per se.[49] 14-3-3 has also been identified in yeast and implicated in regulation of growth.[50]

Using the partially purified sample of REKS, we examined the effect of the pure sample of 14-3-3.[51] REKS activity is synergistically enhanced by a combination of mammalian recombinant GTPγS-Ki-RAS and rat brain 14-3-3. 14-3-3 did not affect MEK or ERK activity. 14-3-3 neither interacted with Ki-RAS nor affected the GAP activity of neurofibromin under the conditions where the N-terminal fragment of c-Raf-1 inhibited it. These results suggest that 14-3-3 has an additional function in the regulation of the RAS-MEK-ERK cascade pathway through the activation of REKS. Independently, it has been reported by other groups that 14-3-3 interacts with c-Raf-1,[52-55] the middle T antigen,[56] Bcr[57] and mitochondrial import stimulating factor.[58,59]

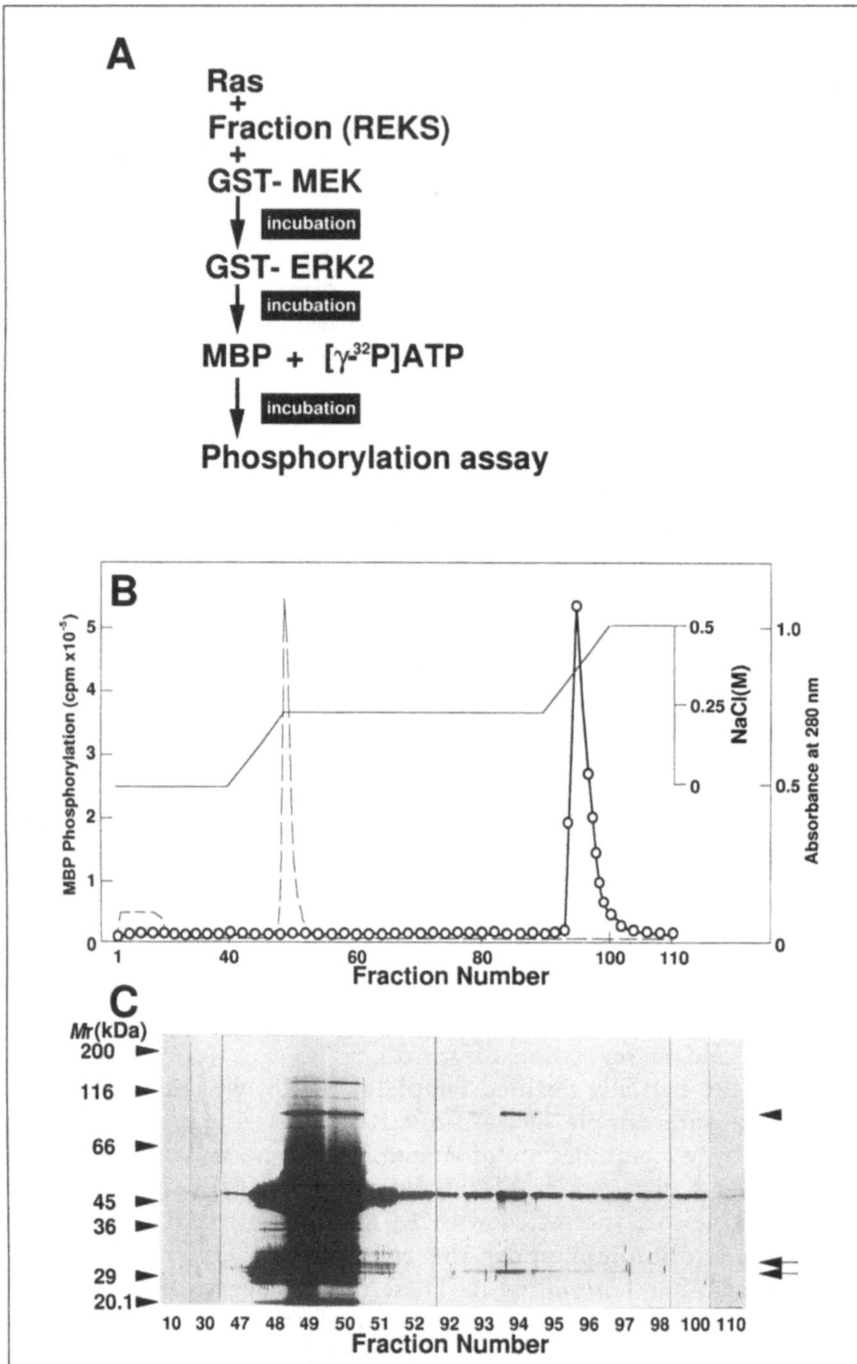
A
Ras
+
Fraction (REKS)
+
GST- MEK
incubation
GST- ERK2
incubation
MBP + [γ-32P]ATP
incubation
Phosphorylation assay
B
MBP Phosphorylation (cpm x10-3)
NaCl(M)
Absorbance at 280 nm
Fraction Number
C
Mr(kDa)
200
116
66
45
36
29
20.1
10 30 47 48 49 50 51 52 92 93 94 95 96 97 98 100 110
Fraction Number

Identification of Bovine Brain REKS as B-Raf Complexed with 14-3-3

We attempted to identify and purify REKS from mammalian tissues and recently succeeded in purifying it to near homogeneity from the cytosol of bovine brain by various combinations of column chromatographies (Fig. 6.2, B and C).[30] It contained three proteins of about 95 kDa, 32 kDa and 30 kDa. Immunoblot analysis revealed that the 95 kDa protein is B-Raf, and the two other proteins belong to the 14-3-3 family. Moreover, the REKS activity was specifically immunoprecipitated by the antibody raised against the C-terminal region of B-Raf. Bovine REKS was activated by the lipid-modified form of GTPγS-Ki-RAS far more effectively than by the lipid-unmodified form. The lipid-modified form of GDP-Ki-RAS was inactive. The K_a value for the lipid-modified form of GTPγS-Ki-RAS is about 20 nM. As described above, 14-3-3 and GTPγS-Ki-RAS synergistically activate *Xenopus* REKS.[51] Moreover, 14-3-3 has been shown to directly interact with c-Raf-1 in in vivo and in vitro systems and enhance its activity.[52-55] The conclusion that bovine REKS is B-Raf complexed with 14-3-3 is consistent with these earlier observations.[52-55] However, it still remains to be clarified whether 14-3-3 is essential for the RAS-dependent activation of B-Raf.

Possible Identification of *Xenopus* Oocyte REKS as B-Raf

The minimum Mr of *Xenopus* REKS is 98 kDa as described above, but it does not crossreact with the anti-B-Raf antibody.[41] However, *Xenopus* REKS and bovine REKS share similar physical and kinetic properties: (i) they adsorb to the GTPγS-GST-Ha-RAS-coupled glutathione agarose column; (ii) they show similar minimum Mrs on SDS-PAGE; and (iii) lipid-modifications of RAS are critical for their activation (See below). These observations suggest that the 98 kDa protein of *Xenopus* REKS is B-Raf. The reason that the 98 kDa protein of *Xenopus* REKS is not recognized by the anti-B-Raf antibody may be the possible difference in the C-terminal amino acid sequence between *Xenopus* and bovine B-Raf.

Fig. 6.2 (opposite). Purification of REKS from bovine brain cytosol. Bovine brain cytosol was subjected to Mono S column chromatography and active fractions were collected. They were subjected to Mono Q column chromatography and active fractions were collected. They were subjected to GTPγS-GST-Ha-RAS coupled glutathione-agarose column chromatography. Elution was performed by reduced glutathione. The active eluate fractions were subjected to a second Mono Q column chromatography. Each fraction (1 μl) of second Mono Q column chromatography was assayed for REKS activity. Another aliquot (60 μl) of each fraction was subjected to SDS-PAGE followed by silver staining. (A) The procedure for the modified cell-free assay system. (B) The REKS activity. (C) Silver staining. An arrowhead and arrows indicate the positions of the 95 kDa protein, and both the 32 and 30 kDa proteins, respectively. The results shown are representative of three independent experiments.

IV. RAF ACTIVATION

Possible Existence of Another B-RAF-activating Factor

The bovine REKS activity immunoprecipitated by the anti-B-Raf antibody is enhanced by the antibody in a GTPγS-Ki-RAS-independent manner.[30] The N-terminal domain of c-Raf-1 inhibits its kinase activity, presumably by masking its kinase domain.[60-62] It is possible that the association of the anti-B-Raf antibody with B-Raf releases it from the inhibitory action of the N-terminal domain, resulting in the activation of B-Raf. This result has raised the possibility that there may be another factor that mimics the action of the antibody.

Inability of RAS to Activate c-RAF-1 in a Cell-free System

A-Raf, B-Raf and c-Raf-1 are expressed in brain.[63] We have identified REKS as B-Raf, but not as A-Raf or c-Raf-1, in bovine brain. On the other hand, c-Raf-1 directly interacts with GTP-RAS in cell-free assay systems[23-25,28] and in yeast two hybrid systems,[26,27] and c-Raf-1 preactivated in intact cells activates MEK in cell-free systems.[8-10] Although these earlier observations suggest that c-Raf-1 activated by GTP-RAS activates MEK, no direct evidence has thus far been obtained that GTP-RAS activates c-Raf-1 or A-Raf in cell-free assay systems. The result that bovine brain A-Raf or c-Raf-1 is not activated by GTPγS-Ki-RAS in our system may be due to the contamination of some interfering materials in the A-Raf or c-Raf-1 fractions, or to the absence of a certain factor necessary for the RAS-dependent A-Raf or c-Raf-1 activation.

Activation of Purified B-Raf by Rap1

The Rap1 family consists of two highly homologous members, Rap1A and Rap1B, and shares overall 50% sequence identity with RAS.[31-33] Moreover, the effector domains of Rap1 and RAS are identical, suggesting that Rap1 is functionally similar and/or antagonistic to RAS. Consistently, Rap1A and/or Rap1B have been shown to antagonize various RAS functions, such as the Ki-RAS-induced transformation of NIH 3T3 cells,[33] the Ha-RAS-induced germinal vesicle breakdown in *Xenopus* oocytes,[64] the N-RAS-inhibited muscarinic K^+ channel activity,[65] the Ki-RAS-induced activation of the *c-fos* promoter/enhancer in NIH 3T3 cells,[66] the proliferation of middle T antigen-transformed Rat-2 cells[67] and the Ha-RAS-induced activation of the MAP kinase cascade in Rat-1 cells,[68] whereas we have shown that Rap1B as well as Ki-RAS stimulates DNA synthesis in Swiss 3T3 cells.[69] Rap1 has been shown to interact with c-Raf-1 in a yeast two hybrid system.[24] Very recently, Rap1A has been cocrystallized with the RAS-binding domain of c-Raf-1.[76] However, no evidence has thus far been obtained that Rap1 directly affects the c-Raf-1 activity.

We examined, using our cell-free assay system, whether Rap1B stimulates B-Raf activity or whether Rap1B antagonizes the Ki-RAS-stimulated B-Raf activity.[34] Recombinant GTPγS-Rap1B stimulates the activity of B-Raf partially purified from bovine brain and immunoprecipitated by the anti-B-Raf antibody. The lipid-modified form of GTPγS-Rap1B was active, but GDP-Rap1B was inactive (Fig. 6.3). Rap1B enhanced the Ki-RAS-stimulated B-Raf activity in an additive manner. These results indicate that both RAS and Rap1 interact with B-Raf and induce its activation in a similar manner, and suggest that Rap1 as well as RAS induces the activation of the B-Raf-dependent MAP kinase cascade in intact cells.

It has been shown in many systems that Rap1 antagonizes various RAS functions, some of which have been considered to be mediated by the action of c-Raf-1.[33,64-68] c-Raf-1 has been shown to be expressed in most tissues, but B-Raf has been shown to be expressed in brain, spinal cord and testis.[63] Our analysis has, however, revealed that B-Raf is also expressed in various rat tissues, suggesting that both c-Raf-1 and B-Raf are expressed in most tissues.[34] It still remains to be clarified why RAS and Rap1 antagonize in the c-Raf-1 activation, while Rap1 as well as RAS activates B-Raf.

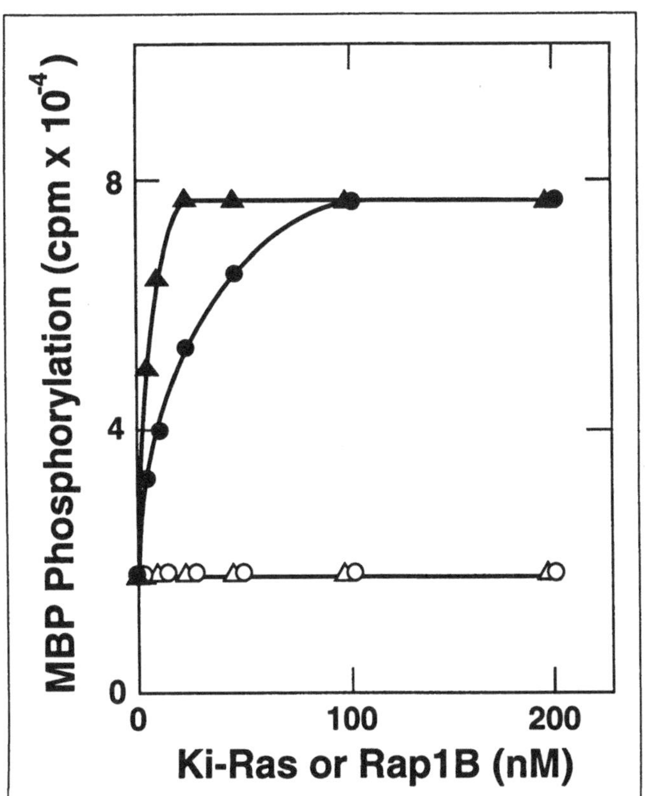

Fig. 6.3. Effect of the lipid-modifications of Ki-RAS and Rap1B on the B-Raf activity. The activity of partially purified B-Raf was measured with various concentrations of lipid-modified or lipid-unmodified forms of GTPγS-Ki-RAS or GTPγS-Rap1B. (▲), with lipid-modified GTPγS-Ki-RAS; (●), with lipid-modified GTPγS-Rap1B; (Δ), with lipid-unmodified GTPγS-Ki-RAS; (O), with lipid-unmodified GTPγS-Rap1B. The results shown are representative of three independent experiments.

Figure 6.4 shows a model for the RAS- and Rap1-dependent regulation of the Raf family. The GDP-bound form of RAS is activated by conversion to the GTP-bound form on the plasma membrane by the action of GDP/GTP exchange proteins such as SOS and Cdc25 (for a review, see ref. 77). SOS is activated by Grb2, which interacts with the tyrosine kinase-type membrane receptors such as the PDGF and EGF receptors (for a review, see ref. 77), whereas Cdc25 is activated by the Ca^{2+}-calmodulin complex in cortical neurons.[78] The GDP-bound form of Rap1B has recently been shown to be converted to the GTP-bound form on the plasma membrane by the action of C3G, which is activated by the action of Crk[79,80] and protein kinase A (see below). Crk is shown to be tyrosine-phosphorylated by c-Abl, BCR/Abl, and the insulin-like growth factor I receptor.[81-83] GTP-RAS activates c-Raf-1 and B-Raf, whereas Rap1 activates B-Raf but antagonizes the RAS-induced activation of c-Raf-1. It is important to establish the cell-free assay system in which RAS stimulates c-Raf-1 activity, and to examine whether Rap1 affects c-Raf-1 activity.

No Effect of the Protein Kinase A-catalyzed Phosphorylation of Rap1B on B-Raf Activation

Rap1A and/or Rap1B have been shown to be phosphorylated by protein kinase A in both intact cells and cell-free systems,[84-89] by

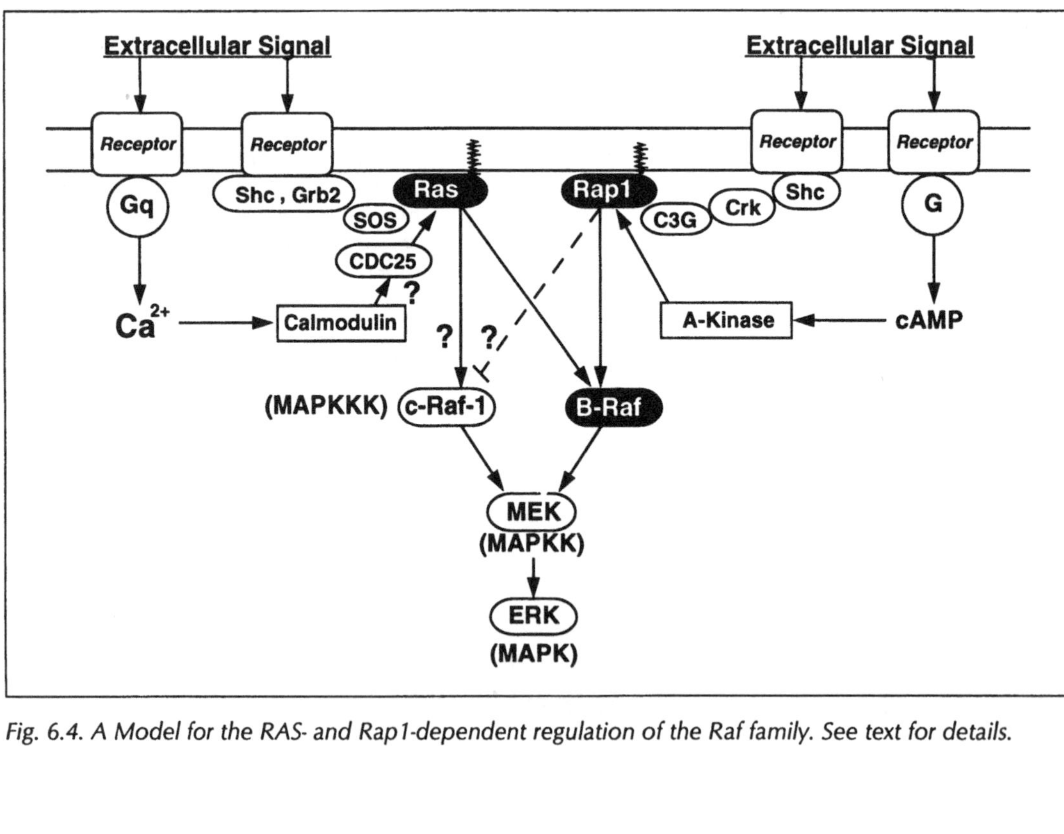

Fig. 6.4. A Model for the RAS- and Rap1-dependent regulation of the Raf family. See text for details.

Ca^{2+}/calmodulin-dependent protein kinase Gr[90] or protein kinase G in a cell-free system.[91] The protein kinase A-catalyzed phosphorylation sites of Rap1A and Rap1B are Ser180 and Ser179, respectively, in their C-terminal regions.[89,92] This phosphorylation of Rap1B lowers its membrane-binding activity and induces its translocation from the membrane to the cytosol.[86,92,93] The phosphorylation of Rap1B makes it sensitive to the action of Smg GDS to stimulate the GDP/GTP exchange reaction.[92] We found that the protein kinase A-catalyzed phosphorylation of Rap1B does not affect B-Raf activation.[34] These results suggest that the protein kinase A-catalyzed phosphorylation of Rap1 contributes to its activation but not to B-Raf activation.

It has been shown that the cyclic AMP-protein kinase A system induces the activation of the MAP kinase cascade in intact PC12 cells[95] and intact COS7 cells.[96] However, it has also been described that the cyclic AMP-protein kinase A system inhibits B-Raf activity in intact PC12 cells[12] and the RAS-dependent MAP kinase activity in intact NIH 3T3 cells and Rat-1 cells,[97] in Rat-1 cells overexpressing human epidermal growth factor receptor[98] and in a cell-free assay system.[99] The reason for this apparent discrepancy is not known, but there are several explanations: for instance, (i) the protein kinase A activates Rap1, whereas it blocks the RAS-induced activation of c-Raf-1 by phosphorylating c-Raf-1 at Ser 43;[98] (ii) the cyclic AMP-protein kinase A system affects the B-Raf activity in a Rap1-independent manner; and (iii) the degrees of these Rap1-dependent and -independent systems affect the B-Raf activity differently depending on the cell types. Further studies are necessary to clarify these possibilities.

Importance of the Lipid-Modifications of RAS and Rap1 for B-Raf Activation

RAS has a Cys-A-A-X motif at its C-terminus (A: aliphatic amino acid; X: any amino acid) which undergoes posttranslational modifications: the cysteine residue is first farnesylated, followed by proteolytic removal of the A-A-X portion and the carboxyl methylation of the exposed cysteine residue (for a review, see ref. 100). Rap1 has a Cys-A-A-Leu motif which similarly undergoes posttranslational modifications except that it is geranylgeranylated.[101] These modifications of RAS are essential for its biological activities, such as malignant transformation and activation of yeast adenlyate cyclase.[100,102] We have also shown that they are biochemically important for RAS activation by its regulators, such as Smg GDS,[39] mCdc25[103] and mSOS.[104] The lipid-modifications of Rap1 are also essential for its inhibition of the Ki-RAS-induced transformation of NIH 3T3 cells[33] and the Ki-RAS-induced activation of the *c-fos* promoter/enhancer in NIH 3T3 cells[66] and for its activation by Smg GDS.[39]

To examine the role of the lipid-modifications of Ki-RAS/Rap1B in B-Raf activation, we employed the lipid-modified and -unmodified form of Ki-RAS/Rap1B purified from the membrane and cytosol fractions of Sf9 insect cells, respectively.[39] The lipid-modified form of GTPγS-Ki-RAS/Rap1B activated brain B-Raf which was partially purified by Mono Q column chromatography, but the lipid-unmodified one was nearly inactive (Fig. 6.3). Thus, the lipid-modifications of both RAS and Rap1 are important for their activation and function. However, how these lipid-modifications contribute to the interaction of RAS and Rap1 with their regulators and target molecules still remains to be clarified.

Different Effects of Various Phospholipids on the Ki-RAS-, Ha-RAS- and Rap1B-Induced B-Raf Activation

Structurally, Raf displays significant similarity to the protein kinase C (PKC) superfamily.[105] The PKC superfamily is classified into three families; the conventional PKC (cPKC), new PKC (nPKC) and atypical PKC (aPKC) families (for a review, see ref. 106). The cPKC family contains two cysteine-rich domains and one Ca^{2+}-dependent phospholipid-binding domain. The former is responsible for its activation by diacylglycerol or tumor-promoting phorbol esters, such as 12-*O*-tetra-decanoyl-phorbol-13-acetate (TPA), in the presence of anionic phospholipid. The latter is responsible for its activation by Ca^{2+} in the presence of anionic phospholipids such as phosphatidylserine (PS). The nPKC family contains two cysteine-rich domains responsible for its activation by diacylglycerol or tumor-promoting phorbol ester in the presence of anionic phospholipids.[106] In contrast, the aPKC family contains only one cysteine-rich domain, but its activity is not affected by diacylglycerol or phorbol ester.[106] PKC-ζ, a member of the aPKC family, has been shown to be enhanced by PS and PI-3,4,5-triphosphate (PIP3).[107] Structurally, the Raf family resembles the aPKC family; it contains only one cysteine-rich domain in the amino-terminal region and a catalytic domain in the carboxyl-terminal region, and lacks the Ca^{2+}-dependent phospholipid-binding domain. Recently, it has been shown that c-Raf-1 binds to PS liposomes, although it does not bind to phorbol esters,[108] and that membrane phospholipid enhances the recombinant c-Raf-1 activity.[109]

We examined phospholipid effects on the RAS- and Rap1-dependent B-Raf activation.[110] Bovine brain PS enhanced the Ki-RAS-stimulated B-Raf activity. Phosphatidic acid (PA) was slightly active, but other phospholipids, such as phosphatidylcholine (PC), phosphatidylethanolamine (PE), phosphatidylinositol (PI), PI-4-monophosphate (PIP), PI-4,5-bisphosphate and PIP3, were inactive. However, none of the above phospholipids affected the Ha-RAS-stimulated B-Raf activity, whereas PI, PS, PE and PA inhibited the Rap1B-stimulated B-Raf activity. PC or PIP did not show any effect on the Rap1B-stimulated B-Raf activity. Synthetic PS with two unsaturated fatty acids, such as 1,2-dioleoyl-PS or 1,2-dilinoleoyl-PS, showed the same effect toward the Ki-RAS- and Rap1B-stimulated B-Raf activities, but synthetic PS with two saturated fatty acids, such as 1,2-distearoyl-PS, was inactive. TPA did not affect the stimulatory or inhibitory effect of PS on the Ki-RAS- and Rap1B-stimulated B-Raf activities. PS did not affect the Ki-RAS-, Ha-RAS- or Rap1B-independent basal B-Raf activity, or the MEK or ERK activity. These results indicate that various phospholipids affect the Ki-RAS-, Ha-RAS and Rap1B-induced B-Raf activation differently by interacting with Ki-RAS or Rap1B, but not with B-Raf alone.

V. CONCLUSION

Using the cell-free assay system which we have developed, we have clarified that both RAS and Rap1 directly interact with B-Raf and activate it. The lipid modifications of RAS and Rap1 are essential to this action. 14-3-3 also interacts with B-Raf and affects the RAS-dependent B-Raf activation. However, a cell-free assay system in which RAS or Rap1 activates c-Raf-1 has not yet been established. It is important to develop this assay system for our understanding of the mode of action of RAS or Rap1 on c-Raf-1 activation.

REFERENCES

1. Davis RJ. The mitogen-activated protein kinase signal transduction pathway. J Biol Chem 1993; 268:14553-14556.
2. Blenis J. Signal transduction via the MAP kinases: proceed at your own RSK. Proc Natl Acad Sci USA 1993; 90:5889-5892.
3. Johnson GL, Vaillancourt RR. Sequential protein kinase reactions controlling cell growth and differentiation. Curr Opin Cell Biol 1994; 6:230-238.
4. Daum G, Eisenmann-Tappe I, Fries H-W et al. The ins and outs of Raf kinases. Trends Biochem Sci 1994; 19:474-480.
5. Zheng C-F, Guan K-L. Activation of MEK family kinases requires phosphorylation of two conserved Ser/Thr residues. EMBO J 1994; 13:1123-1131.
6. Alessi DR, Saito Y, Campbell DG et al. Identification of the sites in MAP kinase kinase-1 phosphorylated by p74Raf-1. EMBO J 1994; 13:1610-1619.
7. Gardner AM, Vaillancourt RR, Lange-Carter CA et al. MEK-1 phosphorylation by MEK kinase, Raf, and mitogen-activated protein kinase: analysis of phosphopeptides and regulation of activity. Mol Biol Cell 1994; 5:193-201.
8. Dent P, Haser W, Haystead TA et al. Activation of mitogen-activated protein kinase kinase by v-Raf in NIH 3T3 cells and in vitro. Science 1992; 257:1404-1407.
9. Howe LR, Leevers SJ, Gómez N et al. Activation of the MAP kinase pathway by the protein kinase raf. Cell 1992; 71: 335-342.
10. Kyriakis JM, App H, Zhang XF et al. Raf-1 activates MAP kinase-kinase.

Nature 1992; 358:417-421.
11. Lange-Carter CA, Johnson GL. RAS-dependent growth factor regulation of MEK kinase in PC12 cells. Science 1994; 265:1458-1461.
12. Vaillancourt RR, Gardner AM, Johnson GL. B-Raf-dependent regulation of the MEK-1/mitogen-activated protein kinase pathway in PC12 cells and regulation by cyclic AMP. Mol Cell Biol 1994; 14:6522-6530.
13. Jaiswal RK, Moodie SA, Wolfman A et al. The mitogen-activated protein kinase cascade is activated by B-Raf in response to nerve growth factor through interaction with p21ras. Mol Cell Biol 1994; 14:6944-6953.
14. Moodie SA, Paris MJ, Kolch W et al. Association of MEK1 with p21ras.GMPPNP is dependent on B-Raf. Mol Cell Biol 1994; 14: 7153-7162.
15. Catling AD, Reuter CW, Cox ME et al. Partial purification of a mitogen-activated protein kinase kinase activator from bovine brain. Identification as B-Raf or a B-Raf-associated activity. J Biol Chem 1994; 269:30014-30021.
16. Nebreda AR, Hunt T. The c-mos proto-oncogene protein kinase turns on and maintains the activity of MAP kinase, but not MPF, in cell-free extracts of Xenopus oocytes and eggs. EMBO J 1993; 12:1979-1986.
17. Posada J, Yew N, Ahn NG et al. Mos stimulates MAP kinase in Xenopus oocytes and activates a MAP kinase kinase in vitro. Mol Cell Biol 1993; 13:2546-2553.
18. Lange-Carter CA, Pleiman CM, Gardner AM et al. A divergence in the MAP kinase regulatory network defined by MEK kinase and Raf. Science 1993; 260:315-319.
19. Kolch W, Heidecker G, Lloyd P et al. Raf-1 protein kinase is required for growth of induced NIH/3T3 cells. Nature 1991; 349:426-428.
20. Dickson B, Sprenger F, Morrison D et al. Raf functions downstream of RAS1 in the Sevenless signal transduction pathway. Nature 1992; 360:600-603.
21. Han M, Golden A, Han Y et al. *C. elegans* lin-45 raf gene participates in let-60 ras-stimulated vulval differentiation. Nature 1993; 363:133-140.
22. Schaap D, van der Wal J, Howe LR et al. A dominant-negative mutant of raf blocks mitogen-activated protein kinase activation by growth factors and oncogenic p21ras. J Biol Chem 1993; 268:20232-20236.
23. Moodie SA, Willumsen BM, Weber MJ et al. Complexes of RAS·GTP with Raf-1 and mitogen-activated protein kinase kinase. Science 1993; 260:1658-1661.
24. Zhang XF, Settleman J, Kyriakis JM et al. Normal and oncogenic p21ras proteins bind to the amino-terminal regulatory domain of c-Raf-1. Nature 1993; 364:308-313.
25. Warne PH, Viciana PR, Downward J. Direct interaction of RAS and the amino-terminal region of Raf-1 in vitro. Nature 1993; 364:352-355.
26. Van Aelst L, Barr M, Marcus S et al. Complex formation between RAS and RAF and other protein kinases. Proc Natl Acad Sci USA 1993; 90:6213-6217.

27. Vojtek AB, Hollenberg SM, Cooper JA. Mammalian RAS interacts directly with the serine/threonine kinase Raf. Cell 1993; 74:205-214.
28. Koide H, Satoh T, Nakafuku M et al. GTP-dependent association of Raf-1 with Ha-RAS: identification of Raf as a target downstream of RAS in mammalian cells. Proc Natl Acad Sci USA 1993; 90:8683-8686.
29. Itoh T, Kaibuchi K, Masuda T et al. A protein factor for ras p21-dependent activation of mitogen-activated protein (MAP) kinase through MAP kinase kinase. Proc Natl Acad Sci USA 1993; 90:975-979.
30. Yamamori B, Kuroda S, Shimizu K et al. Purification of a RAS-dependent mitogen-activated protein kinase kinase kinase from bovine brain cytosol and its identification as a complex of B-Raf and 14-3-3 proteins. J Biol Chem 1995; 2706:11723-11726.
31. Kawata M, Matsui Y, Kondo J et al. A novel small molecular weight GTP-binding protein with the same putative effector domain as the ras proteins in bovine brain membranes. Purification, determination of primary structure, and characterization. J Biol Chem 1988; 263:18965-18971.
32. Pizon V, Chardin P, Lerosey I et al. Human cDNAs rap1 and rap2 homologous to the Drosophila gene Dras3 encode proteins closely related to ras in the 'effector' region. Oncogene 1988; 3:201-204.
33. Kitayama H, Sugimoto Y, Matsuzaki T et al. A ras-related gene with transformation suppressor activity. Cell 1989; 56:77-84.
34. Ohtsuka T, Shimizu K, Yamamori B et al. Activation of brain B-Raf protein kinase by Rap1B small GTP-binding protein. J Biol Chem 1996; 271:1258-1261.
35. Itoh T, Kaibuchi K, Masuda T et al. The post-translational processing of ras p21 is critical for its stimulation of mitogen-activated protein kinase. J Biol Chem 1993; 268:3025-3028.
36. Matsuda S, Kosako H, Takenawa K et al. Xenopus MAP kinase activator: identification and function as a key intermediate in the phosphorylation cascade. EMBO J 1992; 11:973-982.
37. Gotoh Y, Moriyama K, Matsuda S et al. Xenopus M phase MAP kinase: isolation of its cDNA and activation by MPF. EMBO J 1991; 10: 2661-2668.
38. Miyano O, Kameshita I, Fujisawa H. Purification and characterization of a brain-specific multifunctional calmodulin-dependent protein kinase from rat cerebellum. J Biol Chem 1992; 267: 1198-1203.
39. Mizuno T, Kaibuchi K, Yamamoto T et al. A stimulatory GDP/GTP exchange protein for smg p21 is active on the post-translationally processed form of c-Ki-ras p21 and rhoA p21. Proc Natl Acad Sci USA 1991; 88:6442-6446.
40. Sturgill TW, Ray LB, Erikson E et al. Insulin-stimulated MAP-2 kinase phosphorylates and activates ribosomal protein S6 kinase II. Nature 1988; 334:715-718.
41. Kuroda S, Shimizu K, Yamamori B et al. Purification and characterization of REKS from Xenopus eggs. Identification of REKS as a RAS-dependent mitogen-activated kinase kinase kinase. J Biol Chem 1995;

270:2460-2465.
42. Moore BW, Perez VJ. Specific acidic proteins of the nervous system. In: Carlson FD, ed. Physiological and Biochemical aspects of nervous integration, Prentice-Hall, Englewood Cliffs, NJ, 1967: 343-359.
43. Aitken A, Collinge DB, van Heusden BP et al. 14-3-3 proteins: a highly conserved, widespread family of eukaryotic proteins. Trends Biochem Sci 1992; 17:498-501.
44. Boston PF, Jackson P, Kynoch PAM et al. Purification, properties, and immunohistochemical localization of human brain 14-3-3 protein. J Neurochem 1982; 38:1466-1474.
45. Yamauchi T, Nakata H, Fujisawa H. A new activator protein that activates tryptophan 5-monooxygenase and tyrosine 3-monooxygenase in the presence of Ca^{2+}-, calmodulin-dependent protein kinase. Purification and characterization. J Biol Chem 1981; 256:5404-5409.
46. Aitken A, Ellis CA, Harris A et al. Kinase and neurotransmitters. Nature 1990; 344:594.
47. Isobe T, Hiyane Y, Ichimura T et al. Activation of protein kinase C by the 14-3-3 proteins homologous with Exo1 protein that stimulates calcium-dependent exocytosis. FEBS Lett 1992; 308:121-124.
48. Morgan A, Burgoyne RD. Exo1 and Exo2 proteins stimulate calcium-dependent exocytosis in permeabilized adrenal chromaffin cells. Nature 1992; 355:833-836.
49. Zupan LA, Steffens DL, Berry CA et al. Cloning and expression of a human 14-3-3 protein mediating phospholipolysis. Identification of an arachidonoyl-enzyme intermediate during catalysis. J Biol Chem 1992; 267:8707-8710.
50. van Heusden GP, Wenzel TJ, Lagendijk EL et al. Characterization of the yeast BMH1 gene encoding a putative protein homologous to mammalian protein kinase II activators and protein kinase C inhibitors. FEBS Lett 1992; 302:145-150.
51. Shimizu K, Kuroda S, Yamamori B et al. Synergistic activation by RAS and 14-3-3 protein of a mitogen-activated protein kinase kinase kinase named RAS-dependent extracellular signal-regulated kinase kinase stimulator. J Biol Chem 1994; 269:22917-22920.
52. Irie K, Gotoh Y, Yashar BM et al. Stimulatory effects of yeast and mammalian 14-3-3 protein on the Raf protein kinase. Science 1994; 265:1716-1719.
53. Freed E, Symons M, Macdonald SG et al. Binding of 14-3-3 proteins to the protein kinase Raf and effects on its activation. Science 1994; 265:1713-1716.
54. Fu H, Xia K, Pallas DC et al. Interaction of the protein kinase Raf-1 with 14-3-3 proteins. Science 1994; 266:126-129.
55. Fantl WJ, Muslin AJ, Kikuchi A et al. Activation of Raf-1 by 14-3-3 proteins. Nature 1994; 371:612-614.
56. Pallas DC, Fu H, Haehnel LC et al. Association of polyomavirus middle tumor antigen with 14-3-3 proteins. Science 1994; 265:535-537.

57. Reuther GW, Fu H, Cripe LD et al. Association of the protein kinases c-Bcr and Bcr-Abl with proteins of the 14-3-3 family. Science 1994; 266:129-133.
58. Hachiya N, Alam R, Sakasegawa Y et al. A mitochondrial import factor purified from rat liver cytosol is an ATP- dependent conformational modulator for precursor proteins. EMBO J 1993; 12:1579-1586.
59. Hachiya N, Komiya T, Alam R et al. MSF, a novel cytoplasmic chaperone which functions in precursor targeting to mitochondria. EMBO J 1994; 13:5146-5154.
60. Stanton VP, Nichols DW, Laudano AP et al. Definition of the human raf amino-terminal regulatory region by deletion mutagenesis. Mol Cell Biol 1989; 9:639-647.
61. Heidecker G, Huleihel M, Cleveland JL et al. Mutational activation of c-Raf-1 and definition of the minimal transforming sequence. Mol Cell Biol 1990; 10:2503-2512.
62. Bruder JT, Heidecker G, Rapp UR. Serum-, TPA-, and RAS-induced expression from Ap-1/Ets-driven promoters requires Raf-1 kinase. Genes Dev 1992; 6:545-556.
63. Storm SM, Cleveland JL, Rapp UR. Expression of raf family proto-oncogene in normal mouse tissues. Oncogene 1990; 5:345-351.
64. Campa MJ, Chang KJ, Molina y Vedia L et al. Inhibition of ras-induced germinal vesicle breakdown in Xenopus oocytes by rap-1B. Biochem Biophys Res Commun 1991; 174:1-5.
65. Yatani A, Quilliam LA, Brown AM et al. Rap1A antagonizes the ability of RAS and RAS-Gap to inhibit muscarinic K^+ channels. J Biol Chem 1991; 266:22222-22226.
66. Sakoda T, Kaibuchi K, Kishi K et al. smg/rap1/Krev-1 p21s inhibit the signal pathway to the c-fos promoter/enhancer from c-Ki-ras p21 but not from c-Raf-1 kinase in NIH3T3 cells. Oncogene 1992; 7:1705-1711.
67. Jelinek MA, Hassell JA. Reversion of middle T antigen-transformed Rat-2 cells by Krev-1: implications for the role of p21c-ras in polyomavirus-mediated transformation. Oncogene 1992; 7: 1687-1698.
68. Cook SJ, Rubinfeld B, Albert I et al. RapV12 antagonizes RAS-dependent activation of ERK1 and ERK2 by LPA and EGF in Rat-1 fibroblasts. EMBO J 1993; 12:3475-3485.
69. Yoshida Y, Kawata M, Miura Y et al. Microinjection of smg/rap1/Krev-1 p21 into Swiss 3T3 cells induces DNA synthesis and morphological changes. Mol Cell Biol 1992; 12:3407-3414.
70. Omitted in proof.
71. Omitted in proof.
72. Omitted in proof.
73. Omitted in proof.
74. Omitted in proof.
75. Omitted in proof.
76. Nassar N, Horn G, Herrmann C et al. The 2.2 A crystal structure of the RAS-binding domain of the serine/threonine kinase c-Raf-1 in complex

with Rap1A and a GTP analogue. Nature 1995; 375:554-560.
77. McCormick F. Activators and effectors of ras p21 proteins. Curr Opin Genet Dev 1994; 4:71-76.
78. Farnsworth CL, Freshney NW, Rosen LB et al. Calcium activation of RAS mediated by neuronal exchange factor RAS-GRF. Nature 1995; 376:524-527.
79. Tanaka S, Morishita T, Hashimoto Y et al. C3G, a guanine nucleotide-releasing protein expressed ubiquitously, binds to the Src homology 3 domains of CRK and GRB2/ASH proteins. Proc Natl Acad Sci USA 1994; 91:3443-3447.
80. Gotoh T, Hattori S, Nakamura S et al. Identification of Rap1 as a target for the Crk SH3 domain-binding guanine nucleotide-releasing factor C3G. Mol Cell Biol 1995; 15:6746-6753.
81. Feller SM, Knudsen B, Hanafusa H. c-Abl kinase regulates the protein binding activity of c-Crk. EMBO J 1994; 13:2341-2351.
82. ten Hoeve J, Kaarinten V, Fioretos T et al. Cellular interactions of CRKL, and SH2-SH3 adaptor protein. Cancer Res 1994; 54: 2563-2567.
83. Beitner-Johnson D, LeRoith D. Insulin-like growth factor-I stimulates tyrosine phosphorylation of endogenous c-Crk. J Biol Chem 1995; 270: 5187-5190.
84. Hoshijima M, Kikuchi A, Kawata M et al. Phosphorylation by cyclic AMP-dependent protein kinase of a human platelet Mr 22,000 GTP-binding protein (smg p21) having the same putative effector domain as the ras gene products. Biochem Biophys Res Commun 1988; 157:851-860.
85. Kawata M, Kikuchi A, Hoshijima M et al. Phosphorylation of smg p21, a ras p21-like GTP-binding protein, by cyclic AMP-dependent protein kinase in a cell-free system and in response to prostaglandin E1 in intact human platelets. J Biol Chem 1989; 264:15688-15695.
86. Lapetina EG, Lacal JC, Reep BR et al. A ras-related protein is phosphorylated and translocated by agonists that increase cAMP levels in human platelets. Proc Natl Acad Sci USA 1989; 86:3131-3134.
87. Siess W, Winegar DA, Lapetina EG. Rap1-B is phosphorylated by protein kinase A in intact human platelets. Biochem Biophys Res Commun 1990; 170:944-950.
88. Lerosey I, Pizon V, Tavitian A et al. The cAMP-dependent protein kinase phosphorylates the rap1 protein in vitro as well as in intact fibroblasts, but not the closely related rap2 protein. Biochem Biophys Res Commun 1991; 175:430-436.
89. Quilliam LA, Mueller H, Bohl BP et al. Rap1A is a substrate for cyclic AMP-dependent protein kinase in human neutrophils. J Immunol 1991; 147:1628-1635.
90. Sahyoun N, Mcdonald OB, Farrell F et al. Phosphorylation of a RAS-related GTP-binding protein, Rap-1b, by a neuronal Ca^{2+}/calmodulin-dependent protein kinase, CaM kinase Gr. Proc Natl Acad Sci USA 1991; 88:2643-2647.
91. Miura Y, Kaibuchi K, Itoh T et al. Phosphorylation of smg p21B/rap1B

p21 by cyclic GMP-dependent protein kinase. FEBS Lett 1992; 297:171-174.
92. Hata Y, Kaibuchi K, Kawamura S et al. Enhancement of the actions of smg p21 GDP/GTP exchange protein by the protein kinase A-catalyzed phosphorylation of smg p21. J Biol Chem 1991; 266:6571-6577.
93. Lazarowski ER, Lacal JC, Lapetina EG. Agonist-induced phosphorylation of an immunologically ras-related protein in human erythroleukemia cells. Biochem Biophys Res Commun 1989; 161:972-978.
94. Omitted in proof.
95. Frodin M, Peraldi P, Van Obberghen E. Cyclic AMP activates the mitogen-activated protein kinase cascade in PC12 cells. J Biol Chem 1994; 269:6207-6214.
96. Faure M, Voyno-Yasenetskaya TA, Bourne HR. cAMP and beta gamma subunits of heterotrimeric G proteins stimulate the mitogen-activated protein kinase pathway in COS-7 cells. J Biol Chem 1994; 269:7851-7854.
97. Burgering BM, Pronk GJ, wan Weeren PC et al. cAMP antagonizes p21ras-directed activation of extracellular signal-regulated kinase 2 and phosphorylation of mSOS nucleotide exchange factor. EMBO J 1993; 12:4211-4220.
98. Wu J, Dent P, Jelinek T et al. Inhibition of the EGF-activated MAP kinase signaling pathway by adenosine 3',5'-monophosphate. Science 1993; 262:1065-1069.
99. VanRenterghem B, Browning MD, Maller JL. Regulation of mitogen-activated protein kinase activation by protein kinases A and C in a cell-free system. J Biol Chem 1994; 269:24666-24672.
100. Magee AI, Newman CM, Giannakouros T et al. Lipid modifications and function of the ras superfamily of proteins. Biochem Soc Trans 1992; 20:497-499.
101. Kawata M, Farnsworth CC, Yoshida Y et al. Posttranslationally processed structure of human platelet protein smg p21B: Evidence for geranylgeranylation and carboxyl methylation of the C-terminal cysteine. Proc Natl Acad Sci USA 1990; 87:8960-8964.
102. Horiuchi H, Kaibuchi K, Kawamura M et al. The posttranslational processing of ras p21 is critical for its stimulation of yeast adenylate cyclase. Mol Cell Biol 1992; 12:4515-4520.
103. Orita S, Kaibuchi K, Kuroda S et al. Comparison of kinetic properties between two mammalian ras p21 GDP/GTP exchange proteins, ras guanine nucleotide-releasing factor and smg GDP dissociation stimulator. J Biol Chem 1993; 268:25542-25546.
104. Nakanishi H, Orita S, Kaibuchi K et al. Kinetic properties of Ash/Grb2-interacting GDP/GTP exchange protein. Biochem Biophys Res Commun 1994; 198:1255-1261.
105. Kazanietz MG, Bustelo XR, Barbacid M et al. Zinc finger domains and phorbol ester pharmacophore. Analysis of binding to mutated form of protein kinase C zeta and the vav and c-Raf proto-oncogene products. J Biol Chem 1994; 269:11590-11594.

106. Nishizuka Y. Intracellular signaling by hydrolysis of phospholipids and activation of protein kinase C. Science 1992; 258:607-614.
107. Nakanishi H, Brewer KA, Exton JH. Activation of the zeta isozyme of protein kinase C by phosphatidylinositol 3,4,5-triphosphate. J Biol Chem 1993; 268:13-16.
108. Ghosh S, Xie WQ, Quest AF et al. The cysteine-rich region of Raf-1 kinase contains Zinc, translocates to liposomes, and is adjacent to a segment that binds to GTP-RAS. J Biol Chem 1994; 269:10000-10007.
109. Dent P, Reardon DB, Morrison DK et al. Regulation of Raf-1 and Raf-1 mutants by RAS-dependent and RAS-independent mechanisms in vitro. Mol Cell Biol 1995; 15:4125-4135.
110. Kuroda S, Ohtsuka T, Yamamori B et al. Different effects of various phospholipids on Ki-Ras-, Ha-Ras, and Rap1B-induced B-Raf activation. J Biol Chem 1996; 271:14682-14683.

INDEX

Kolibri 360 GmbH
Ferdinand-Jühlke-Straße 7
99095 Erfurt, Deutschland
produktsicherheit@kolibri360.de